Lecture Notes in Computer Science 769

Edited by G. Goos and J. Hartmanis

J. L. Nazareth

The Newton-Cauchy Framework

A Unified Approach to
Unconstrained Nonlinear Minimization

Springer-Verlag
Berlin Heidelberg New York
London Paris Tokyo
Hong Kong Barcelona
Budapest

J. L. Nazareth

The Newton-Cauchy Framework

A Unified Approach to
Unconstrained Nonlinear Minimization

Springer-Verlag

Berlin Heidelberg New York
London Paris Tokyo
Hong Kong Barcelona
Budapest

Series Editors

Gerhard Goos
Universität Karlsruhe
Postfach 69 80
Vincenz-Priessnitz-Straße 1
D-76131 Karlsruhe, Germany

Juris Hartmanis
Cornell University
Department of Computer Science
4130 Upson Hall
Ithaca, NY 14853, USA

Author

J. L. Nazareth
Department of Pure and Applied Mathematics, Washington State University
Pullman, WA 99164-3113, USA

CR Subject Classification (1991): G.1.6, J.1, J.2, J.6
1991 Mathematics Subject Classification: 49Mxx, 65Kxx, 90-08, 90Cxx

ISBN 3-540-57671-1 Springer-Verlag Berlin Heidelberg New York
ISBN 0-387-57671-1 Springer-Verlag New York Berlin Heidelberg

Typesetting: Camera-ready by author
45/3140-543210 - Printed on acid-free paper

Preface

Minimizing a nonlinear, multidimensional function $f(x)$, $x \in \Re^n$, where f is smooth and not necessarily convex, is a central problem of computational optimization. An understanding of the methods used to solve it is essential for anyone interested in computation for several reasons:

- Nonlinear minimization problems frequently arise in practice, their solutions either being of immediate interest or required at intermediate stages of a more complex calculation.

- Closely related problems, in particular, solving systems of nonlinear equations or nonlinear least squares data-fitting problems, can be posed as nonlinear minimization problems; alternatively, techniques used in minimization methods can be suitably adapted or specialized to solve such problems.

- Nonlinear ordinary differential equations, partial differential equations or optimal control problems, which are defined over function spaces, must eventually be discretized for solution on a computer. This leads, in turn, to finite-dimensional nonlinear equation-solving or minimization problems.

- Unconstrained minimization techniques form the backbone of methods for solving *constrained* minimization problems. In this regard, it is worth noting that the addition of constraints can *sometimes* render an optimization problem computationally *more* tractable. As an extreme case, suppose f is to be minimized subject to a set of $n - 1$ independent linear equality constraints along with finite lower and upper bounds on the n variables. Then the problem is equivalent to a unidimensional minimization on a line segment.

- Recent interior-point techniques for linear programming (LP), which now supplement Dantzig's simplex method, have moved computational LP away from its traditional base in combinatorial programming, and repositioned it, alongside linearly constrained nonlinear optimization, in the transition region between unconstrained nonlinear minimization on the one hand and nonlinearly constrained optimization on the other. Again, techniques of unconstrained minimization or related techniques of nonlinear equation-solving play a key role in these new interior-point LP methods.

Two classical methods for minimizing a nonlinear function are *Cauchy's method*, which uses a search direction of steepest descent, and *Newton's method*, which uses a search direction derived from a local quadratic approximating model obtained, in turn, from the Taylor expansion. More recently, during the

v

digital computer era, there have been two further algorithmic breakthroughs. The *conjugate gradient method (CG)*, proposed by Hestenes and Stiefel in 1952, and the *variable metric method*, developed by Davidon during the years 1955-1959. The CG method was originally proposed for minimizing a strictly convex quadratic function or, equivalently, for solving a positive definite symmetric system of linear equations, and it was straightforwardly adapted to general nonlinear minimization by Fletcher and Reeves in 1964. Nowadays, CG-related is a generic name for a class of methods that require limited computer storage. The variable metric method was clarified and brought to the attention of the optimization community by Fletcher and Powell in 1963, and subsequently broadened and relabelled under the generic name *quasi-Newton*, incorporating the contributions of numerous researchers. Interestingly enough, Davidon's original paper developed the seminal underlying ideas in the setting of non-quadratic problems, but the important clarification and promulgation of Fletcher and Powell, which gained wide acceptance for the variable metric method, placed considerable emphasis on its theoretical properties on quadratic functions. Thus the historical development of *both* modern breakthroughs of computational nonlinear optimization occurred at the interface with computational linear algebra. More recently, computational nonlinear minimization has broken loose from these early moorings, and its methods can now be formulated quite independently of the historical context.

The literature on unconstrained nonlinear minimization is vast and there are several useful expository texts that discuss individual methods in detail, including, for example, Avriel [1976], Bazaraa, Sherali and Shetty [1993], Dennis and Schnabel [1983], Fletcher [1980], Gill, Murray and Wright [1981] and Luenberger [1984]. *Noticeably lacking, however, is a treatment that reveals the essential unity of the subject.* This is the central concern of our research monograph, namely, to explore the relationships between the main methods, to develop a unifying Newton/Cauchy framework and to point out its rich wealth of algorithmic implications. The monograph also makes a contribution to clarifying the notation of the subject, currently full of conflicts and contradictions, as well as the terminology of quasi-Newton methods, which currently resembles 'alphabet soup', to quote Dennis and Schnabel [1983]. We concentrate on *basic conceptual methods* rather than on algorithmic variants and implementational details, and in the interests of brevity, we try to keep to a minimum the duplication of material that is widely available in the literature and the texts previously cited.

The monograph consists of six main chapters organized as follows:

Chapter 1 introduces the conjugate gradient and quasi-Newton methods within the context of their historical development, namely, convex quadratic minimization. The methods assume a particularly elegant form in this setting and have interesting algebraic properties. To neglect this aspect is to turn one's back on some of most attractive results of the subject, as well as the important interface with computational linear algebra. Our development follows a logical and orderly progression of ideas, highlighting the small set of basic principles that give rise to the methods. However, we do not then, according to historical precedent, generalize the methods in patchwork fashion so that they can be applied to nonquadratic problems. Our objectives in this chapter are much more limited, namely, *to motivate the methods and demonstrate their properties in a particular simple setting.*

In Chapter 2, we begin afresh in the setting of general nonlinear functions. Our starting point for development is the classical steepest-descent *metric-based* method of Cauchy. We follow an orderly progression of ideas leading, in turn, to Davidon's variable metric, the conjugate-gradient metric and the Newton metric, and highlighting the relationship between them.

In Chapter 3, our starting point for development is the classical *model-based* Newton's method. Again, a logical train of ideas lead, in succession, to the quasi-Newton model, the CG-related model and the Cauchy model.

Computational unconstrained nonlinear optimization comes to life from a study of the interplay between the metric-based (Chapter 2) and model-based (Chapter 3) points of view, with the motivating development of Chapter 1 in the background to lend added dimension. This is the topic of Chapter 4, which ties together the preceding three chapters, develops the Newton/Cauchy framework and indicates its rich array of algorithmic implications.

Chapter 5 discusses the basic conditions for establishing global convergence of *implementable algorithms* derived from the Newton/Cauchy framework and the important idea of *hierarchical implementation* of optimization methods.

Finally, Chapter 6 overviews nonlinear unconstrained optimization technology, namely, the wide variety of implementable mathematical and numerical algorithms that can be derived from the basic conceptual methods of the Newton/Cauchy framework.

Chapters 1, 2 and 3 are each largely self-contained and develop all the necessary expository detail. On the other hand, Chapters 4, 5 and 6 are more concise and collectively provide a free-standing structured guide to the unconstrained optimization literature, to which they make extensive reference. They also highlight some important unexplored territory.

The monograph assumes the following background on the part of the reader:

- basic multivariate calculus; see, for example, Rudin [1976],

- the basic factorizations of computational linear algebra; see, for example, Golub and Van Loan [1989] or Watkins [1991],

- elementary convex analysis, in particular, the characterization of convex sets and differentiable convex functions; see, for example, Avriel [1976] or Luenberger [1984],

- the basic algorithms for *one-dimensional* minimization and solving a non-linear equation of a *single* variable; see any introductory numerical analysis text, for example, Kahaner, Moler and Nash [1989],

- and the basic theoretical characterization of optimal points, in particular, definitions of local and global optima, necessary and sufficient optimality conditions for smooth unconstrained functions and some basic exposure to the Lagrangian/Karush-Kuhn-Tucker optimality conditions for constrained optimization. These topics are widely treated in several excellent texts, for example, Avriel [1976], Luenberger [1984], or Zangwill [1969], and their duplication here would be pointless. In this regard, it is also worth noting explicitly that as a result of the repositioning of linear programming mentioned previously, it is increasingly likely that newcomers to optimization algorithms will be introduced to the subject via the methods of unconstrained nonlinear minimization rather than the more traditional combinatorial-based linear programming and the simplex method, and that the next generation of textbooks on computational linear and nonlinear optimization will adopt the following pattern of exposition: 1. Optimality conditions for both unconstrained and constrained optimization. 2. Unconstrained minimization methods. 3. Methods for linearly constrained problems including linear programming. 4. Methods for nonlinearly constrained problems.

The monograph is addressed to a broad spectrum of practitioners, researchers, instructors, and students, and we hope that it proves to be both useful and a refreshing new prespective on computational nonlinear optimization. For pedagogical purposes, it can be used to *supplement* one of the optimization texts cited earlier. It can also be used as the primary text of a graduate research seminar course when the instructor fills in background material as needed, fleshes out Chapters 4 through 6 through assigned readings in the optimization literature, and encourages students to explore uncharted territory via appropriate research projects.

Finally, it is a pleasure to thank various people who have contributed directly or indirectly to this effort. My thanks to Bill Davidon, whose algorithmic genius has always been a source of inspiration. I thank my Optimization Group colleagues Bob Mifflin and Kuruppu Ariyawansa with whom I share many research interests, and also Mike Kallaher, former chairman of WSU's Department of Pure and Applied (and, in all but name, Computational) Mathematics, whose enlightened approach has given the department a secure balance on the three equal legs of the mathematical tripod. My affiliation at the University of Washington has been invaluable, and I have profited from interaction, in particular, with Jim Burke, Alan Goldstein, Bob O'Malley, Terry Rockafellar and Paul Tseng. I thank Brian Smith, Laura Kustiner, and Miguel Gomez - students in the Fall '92 graduate course on computational nonlinear optimization, who provided useful feedback when some of this material was formulated and presented.

And finally, on a personal note, I thank my wife Abigail for her encouragement when the task of writing grew burdensome, as it always does, even for a relatively short monograph such as this.

Bainbridge Is., WA, 1993   J.L. Nazareth

Table of Contents

CHAPTER 1: MOTIVATION

1 Introduction

In this chapter, we motivate the two key modern algorithmic developments of unconstrained nonlinear minimization, namely, the *conjugate gradient method* and the *quasi-Newton method*, within a specific context: the minimization of a strictly convex quadratic function or, equivalently, the solution of a positive-definite symmetric system of linear equations. In this simplified setting, the methods assume an especially elegant form.

Our derivation is *inductive* and follows an orderly progression of ideas intended to highlight the relationships between the conjugate gradient and quasi-Newton methods and their interesting algebraic properties on convex quadratic functions. However, we do not follow historical precedent and generalize the methods piecemeal so that they can then be applied to the minimization of arbitrary smooth functions. Our objectives in this chapter are much more limited, namely, we seek to introduce the two key methods under consideration at their interface with computational linear algebra, and thus provide *background and motivation* for the more general, self-contained formulations of Chapters 2 and 3.

2 Preliminaries

2.1 Quadratic Functions

Consider the problem of minimizing a strictly convex quadratic function $f : \Re^n \to \Re^1$ of the form

$$f(x) = -b^T x + \frac{1}{2} x^T A x, \tag{1}$$

where $b \in \Re^n$ is a given vector and $A \in \Re^{n \times n}$ is a given positive-definite symmetric matrix (henceforth written $A > 0$).

Denote the gradient mapping of f by $\nabla f : \Re^n \to \Re^n$, and the *gradient vector* at the point $x \in \Re^n$ by g. Thus,

$$g = \nabla f(x) = -b + Ax. \tag{2}$$

The matrix $A = \nabla^2 f(x) = [\partial^2 f(x)/\partial x_i \partial x_j]$ is the *Hessian matrix* or matrix of second partial derivatives of f. For a quadratic function, the Hessian does not vary with x, i.e., it is a constant matrix.

Consider any two distinct points, say, x and x_+, with corresponding gradient vectors given by (2), i.e., $g = -b + Ax$ and $g_+ = -b + Ax_+$. Then

$$A(x_+ - x) = g_+ - g. \tag{3}$$

Define the step $s = x_+ - x$ and the corresponding gradient change $y = g_+ - g$, so (3) can be written more conveniently as

$$As = y. \tag{4}$$

At the minimizing point x^* of f, the gradient vector is the zero vector and thus

$$x^* = A^{-1}b = x - A^{-1}g. \tag{5}$$

The vector

$$x^* - x = -A^{-1}g \tag{6}$$

defines the *Newton direction* at the point x and it corresponds to the step from x to the minimizing point x^* of f.

An alternative expression $\bar{f}(x)$ for the convex quadratic function (1), which differs from $f(x)$ by a constant, is

$$\bar{f}(x) = \frac{1}{2}(x - x^*)^T A(x - x^*). \tag{7}$$

It is useful to note that the quantity g can be viewed equivalently as the *residual vector* of the linear system $Ax^* = b$ at a point x that approximates the solution x^*. Throughout this chapter, one can substitute the word "residual" for "gradient" in order to obtain the formulation and terminology more common to computational linear algebra. As already stated, the methods as formulated in this chapter lie at the interface of unconstrained nonlinear optimization and computational linear algebra, i.e., linear equation solving by iterative methods, but there are differences in the way problem data is handled in these two cases that are worth identifying. In the *linear equation solving by iterative methods* setting, the matrix A is usually assumed to be *explicitly* available, but it must not be factorized as happens when the linear system $Ax^* = b$ is solved by a *direct* method. Rather, A is used to compute a residual vector $Ax - b$ at an approximate solution x via a matrix-vector product operation. In contrast, in the *nonlinear optimization* setting, it is generally assumed that the Hessian matrix A of the quadratic function f is only available *implicitly* and it is the gradient vector g that is available explicitly. This may appear a little artificial in the case of a quadratic function, i.e., it would seem that in order to obtain g one must have access to A. However, recall that the more general optimization problem being addressed is that of minimizing an *arbitrary* smooth nonlinear function. Information about this function and its derivatives is provided via an evaluation subroutine within which the Hessian matrix is usually only *implicit*. Given a particular point x, this subroutine returns the function value and gradient vector at x and sometimes, but much less commonly, the Hessian matrix at x. The assumption throughout this first chapter is that the matrix A exists and can be used as a *mathematical operator* for purposes of formulating our methods, but it is *not* available for *computational* usage in algorithms as they might eventually be implemented on a computer. The difference in the way that information about the problem to be solved is provided and handled is a watershed that separates computational linear algebra from nonlinear minimization and distinguishes our methods as they are realized in these two problem settings. We shall return to these issues repeatedly in this chapter and complete this detailed discussion in the concluding Section 11.

2.2   Line Search and Step Length Procedures

Given an approximation x to the minimizing point x^* of (1) and a non-zero direction vector $d \in \Re^n$, the *restriction* of f to the line $x + \alpha d$, $\alpha \in \Re^1$ is defined to be the following function $\rho : \Re^1 \to \Re^1$,

$$\rho(\alpha) = f(x + \alpha d), \quad \alpha \in \Re^1. \tag{8}$$

The derivative of ρ for any value α of its argument is the *directional derivative* of f at the corresponding point $x + \alpha d$, i.e.,

$$\rho'(\alpha) = \nabla f(x + \alpha d)^T d. \tag{9}$$

In particular, $\rho'(0) = g^T d$. If $g^T d < 0$, then d is called a *direction of descent*.

Suppose that ρ is *minimized*, i.e., f is minimized along the direction d. A procedure for finding the minimizing point, say α^*, is called an *exact line search*, and

$$\rho'(\alpha^*) = \nabla f(x + \alpha^* d)^T d = 0. \tag{10}$$

Let $x_+ = x + \alpha^* d$ and $g_+ = \nabla f(x_+)$. Then (10) can be written $g_+^T d = 0$. Also, using (3), $g_+ = g + \alpha^* A d$. These two relations imply that

$$\alpha^* = -\frac{g^T d}{d^T A d}. \tag{11}$$

Since $A > 0$ and $d \neq 0$, it follows that $d^T A d > 0$. Note that the sign of α^* is positive when d is a direction of descent and nonpositive otherwise.

If A is not explicitly available (as discussed at the end of the previous section) and must be accessed indirectly via gradient vectors or function values at points in the domain of f, then evaluate the gradient vector at an intermediate point, say, $x_{1/2}$, i.e.,

$$x_{1/2} = x + \alpha d, \quad g_{1/2} = \nabla f(x_{1/2}), \tag{12}$$

where $\alpha \neq 0$ (typically, $\alpha = 1$), and reformulate expression (11) as follows:

$$\alpha^* = -\frac{g^T(\alpha d)}{d^T A(\alpha d)} = -\frac{g^T(x_{1/2} - x)}{d^T A(x_{1/2} - x)} = -\frac{g^T(x_{1/2} - x)}{d^T(g_{1/2} - g)}. \tag{13}$$

The last equality follows from the analogue of (3). It is possible to circumvent the additional gradient evaluation $g_{1/2}$ at the intermediate point $x_{1/2}$ by observing that its purpose in expression (13) is to furnish the directional derivative

$d^T g_{1/2}$. This could be obtained, alternatively, from two function evaluations and use of the following forward difference approximation:

$$d^T g_{1/2} = [f(x_{1/2} + hd) - f(x_{1/2})]/h, \qquad (14)$$

where h is a small positive number, for example, $h = 2^{-\frac{t}{2}}$ on a computer with t binary-digit floating-point arithmetic. This alternative will be useful when a function value at any point is much cheaper to compute than the corresponding gradient vector.

A procedure for finding an *improving*, but not necessarily minimizing point along d, will be called a *line search*. When a step is taken without regard to whether or not the function value is reduced, we will use the term *step length procedure* in place of line search procedure.

2.3 Sequential Step Length and Sequential Line Search Procedures

Now, instead of a single direction d, consider a given set of linearly independent (hence non-zero) direction vectors, say, $d_1, \ldots, d_k$, $k \leq n$, and the following iteration:

Given a starting point x_1:

for $j = 1, \ldots, k$
$\qquad x_{j+1} = x_j + \alpha_j d_j$, $\alpha_j \in R^1$
end

This is called a *sequential step length procedure*. Denote the gradient at x_j by g_j and note that α_j can be of *either* sign and the function values at x_j need *not* form a monotonic sequence.

When the step length along each direction is chosen so that the function f is minimized, i.e.,

$$g_{j+1}^T d_j = 0, \ \ 1 \leq j \leq k, \qquad (15)$$

then the foregoing procedure is called a *sequential exact line search*. Denote the corresponding step lengths by α_j^*. Then from (11),

$$\alpha_j^* = -\frac{g_j^T d_j}{d_j^T A d_j}, \ \ 1 \leq j \leq k. \qquad (16)$$

If d_j is a direction of descent, i.e., $g_j^T d_j < 0$, then $\alpha_j^* > 0$.

Analogously to (13), expression (16) can be reformulated as follows:

$$\alpha_j^* = -\frac{g_j^T (x_{j+1/2} - x_j)}{d_j^T (g_{j+1/2} - g_j)} \qquad (17)$$

4

where $g_{j+1/2}$ is the gradient vector at an intermediate point $x_{j+1/2} = x_j + \alpha_j d_j$, where α_j is a non-zero step (typically, $\alpha_j = 1$). Again, the explicit evaluation of $g_{j+1/2}$ can be avoided by estimating a directional derivative analogously to (14).

When the function is reduced, but not necessarily minimized at each iteration of the foregoing proecedure, then it is called a *sequential line search*.

3   Conjugate Directions

3.1   Definition and Properties

Next, impose a further restriction on the non-zero directions $d_1, \ldots, d_k$ used in the procedure of Section 2.3, namely, that they satisfy the conditions

$$d_i^T A d_j = 0, \ i \neq j, \ 1 \leq j \leq k. \tag{18}$$

Such directions are said to be mutually *conjugate* with respect to the matrix A, i.e., they are orthogonal with respect to the inner product defined by the positive definite symmetric matrix A. Obviously, the directions d_i are linearly independent.

When conjugate directions are used in a sequential exact line search then the relation (15) enlarges to

$$g_{j+1}^T d_i = 0, \ \ 1 \leq i \leq j \leq k. \tag{19}$$

The proof of (19) is a simple induction as follows:

Make the induction hypothesis that the result (19) is true for a sequential exact line search along mutually conjugate directions $d_1, d_2, \ldots, d_{\kappa-1}$ for any κ, $1 < \kappa \leq k$, i.e.,

$$g_{j+1}^T d_i = 0, \ \ 1 \leq i \leq j \leq \kappa - 1. \tag{20}$$

We must show that the hypothesis is then true for conjugate directions $d_1, d_2, \ldots, d_\kappa$. Since the line search along the last direction d_κ is exact,

$$g_{\kappa+1}^T d_\kappa = 0. \tag{21}$$

Furthermore, for $i \leq \kappa - 1$,

$$
\begin{aligned}
g_{\kappa+1}^T d_i &= (g_\kappa + \alpha_\kappa^* A d_\kappa)^T d_i \ \text{using a version of (3)} \\
&= g_\kappa^T d_i + \alpha_\kappa^* d_\kappa^T A d_i \\
&= 0,
\end{aligned}
\tag{22}
$$

where the last equality follows from the induction hypothesis (20) and mutual conjugacy of the search directions.

5

Thus, combining (20), (21) and (22) gives

$$g_{j+1}^T d_i = 0, \quad 1 \le i \le j \le \kappa. \tag{23}$$

The induction hypothesis is obviously true for $\kappa = 2$ and the result follows. $\square$

The result (19) implies that x_{j+1} is the minimizing point of f in the affine space $\{x_1 + \text{span}\{d_1, \ldots, d_j\}\}$. In particular, observe when $k = n$ that g_{n+1} is orthogonal to n linearly independent directions and thus $g_{n+1} = 0$ and $x_{n+1} = x^*$. Thus a sequential exact line search along n mutually conjugate directions yields the minimizing point x^* in no more than n steps. Of course, it is possible for the gradient to vanish prematurely. For example, when d_1 is along the Newton direction $-A^{-1} g_1$ at x_1, the sequential exact line search finds the minimizing point after a single step.

3.2 Generating Conjugate Directions via Gram-Schmidt Orthogonalization

As previously noted, conjugate directions are simply directions that are orthogonal with respect to the inner product defined by a positive definite symmetric matrix A. Thus the well-known Gram-Schmidt orthogonalization procedure can be used to obtain k conjugate direction vectors from any given set of $k \le n$ linearly independent vectors, say, $v_1, \ldots, v_k$, as follows:

Procedure GS:

$d_1 = v_1$
for $j = 2, \ldots, k$
$\quad d_j = v_j - \sum_{i=1}^{j-1} \left(\frac{d_i^T A v_j}{d_i^T A d_i} \right) d_i$

$$\tag{24}$$

end

At iteration j of this procedure, the vector d_j is chosen to lie in the subspace spanned by v_j and the set of mutually conjugate directions $d_1, \ldots, d_{j-1}$ obtained at previous iterations. Note that

$$\text{span}\{d_1, \ldots, d_{j-1}\} = \text{span}\{v_1, \ldots, v_{j-1}\}.$$

The constants

$$r_{ij} = \frac{d_i^T A v_j}{d_i^T A d_i}, \quad i = 1, \ldots, j-1 \tag{25}$$

in the foregoing expression (24), which define the chosen linear combination of previous directions, ensure that d_j is conjugate (A-orthogonal) to $d_1, \ldots, d_{j-1}$. This can be verified directly.

In light of the discussion at the end of Section 2.1, the constants r_{ij} in (25) must be reexpressed as follows using (3), in order to obtain an implementable version of *Procedure GS*.

$$r_{ij} = \frac{\alpha_i^* d_i^T A v_j}{\alpha_i^* d_i^T A d_i} = \frac{(x_{i+1} - x_i)^T A v_j}{(x_{i+1} - x_i)^T A d_i} = \frac{y_i^T v_j}{y_i^T d_i}, \quad i = 1, \ldots, j - 1, \qquad (26)$$

where $y_i = g_{i+1} - g_i$. Thus (24) can be rewritten for computational purposes as

$$d_j = v_j - \sum_{i=1}^{j-1} \left(\frac{y_i^T v_j}{y_i^T d_i} \right) d_i. \qquad (27)$$

Let us define the $n \times k$ matrices V and D as follows:

$$V = [v_1, \ldots, v_k], \quad D = [d_1, \ldots, d_k],$$

and let R be a $k \times k$ *unit* upper triangular matrix, i.e., $r_{ii} = 1, i = 1, \ldots, k$ with supradiagonal elements given by (26). Then the Gram-Schmidt A-orthogonalization procedure yields the matrix factorization

$$V = DR. \qquad (28)$$

The matrix R is invertible and its inverse R^{-1} is also unit upper triangular, i.e,

$$D = VR^{-1}. \qquad (29)$$

Clearly for each index $j, 1 \leq j \leq k$, the vector d_j is a linear combination of $v_1, \ldots v_j$.

4 The Conjugate Gradient Method

The conjugate gradient method is obtained by making a particular choice for the vectors v_j within the setting of a sequential exact line search procedure.

4.1 A Specialized Gram-Schmidt Procedure

Given a set of n linearly independent vectors $v_1, \ldots, v_n$ and a starting point x_1, let us *interleave* iterations of the foregoing Procedure GS, for generating conjugate vectors d_j, with iterations of a sequential exact line search procedure (Section 2.3) using these as search directions. (Note that the corresponding step lengths α_j^* can be of either sign.) This will yield the minimizing point x^* in, at most, n steps and will be termed a conjugate direction procedure.

Next, let us not preordain the vectors v_j. Instead, let us take $v_1 = -g_1$, the negative gradient at the initial point x_1, which in turn gives $d_1 = -g_1$. Thereafter, let us take

$$v_j = -g_j, \qquad (30)$$

the negative gradient vector obtained from the most recent iterate of the sequential exact line search. Thus, we seek to develop conjugate directions using the (negative) gradient vectors obtained from the sequential exact line search procedure.

Observe that the iterate x_j is the result of a sequential exact line search along conjugate directions $d_1, \ldots, d_{j-1}$. Thus, from Section 3.1, the corresponding gradient vector g_j is orthogonal to $d_1, \ldots, d_{j-1}$. When the vector $g_j \neq 0$, it is then obviously linearly independent of the directions $d_1, \ldots d_{j-1}$ and thus Procedure GS can be continued to yield another conjugate direction. We know that the exact line search procedure must terminate after, at most, n steps. Let $k \leq n$ be the first index for which $g_{k+1} = 0$ and let

$$G = [g_1, \ldots, g_k], \quad D = [d_1, \ldots d_k]. \tag{31}$$

Then expressions (28) and (29) become

$$-G = DR, \quad D = -GR^{-1}, \tag{32}$$

where the supradiagonal elements of the unit upper triangular matrix R are obtained from (26), i.e.,

$$r_{ij} = -\frac{y_i^T g_j}{y_i^T d_i}, \quad 1 \leq i < j \leq k. \tag{33}$$

4.2 A Conjugate Gradient Procedure

A dramatic *simplification* occurs as a result of the foregoing choices. From (32), observe that

$$\text{span}\{d_1, \ldots, d_{j-1}\} = \text{span}\{g_1, \ldots, g_{j-1}\}, \quad 1 < j \leq k+1.$$

Thus g_j is also orthogonal to $g_1, \ldots, g_{j-1}$, and it follows that g_j is orthogonal to $y_1, \ldots, y_{j-2}$, where $y_i = g_{i+1} - g_i \ \forall \ i$. Hence from (33),

$$r_{ij} = 0, \quad i \leq j - 2. \tag{34}$$

Thus expression (27) reduces to

$$d_j = -g_j + \left(\frac{y_{j-1}^T g_j}{y_{j-1}^T d_{j-1}} \right) d_{j-1}. \tag{35}$$

Since line searches are exact, $g_j^T d_{j-1} = 0$ and thus $g_j^T d_j = -g_j^T g_j < 0$. Therefore d_j is a *direction of descent*, justifying the choice of *negative* sign for the gradient vector in (30). If an exact line search is conducted along d_j then we know the corresponding step length α_j^* defined by (16) or (17) is positive.

8

Expression (35) is the conjugate gradient direction originally proposed by Hestenes and Stiefel [1952] and its use within a sequential exact line search procedure yields the *Conjugate Gradient Method* in its most basic form.

We can *summarize* the foregoing development very succinctly as follows: *each gradient vector g_j is conjugate to $d_1, \ldots, d_{j-2}$, i.e., orthogonal to $y_1, \ldots, y_{j-2}$. Thus, in order to obtain the vector d_j, take a linear combination in the span $\{-g_j, d_{j-1}\}$ that is conjugate to d_{j-1}, i.e., orthogonal to y_{j-1}. This vector is given by (35) and, by construction, d_j is then conjugate to all previous directions $d_1, \ldots, d_{j-1}$.*

The procedure, stripped bare of detail, for example, tests to detect premature termination, can be summarized as follows:

Procedure CG:

Given x_1 and associated gradient g_1:
for $j = 1, \ldots, n$

$$d_j = -g_j + \left(\frac{y_{j-1}^T g_j}{y_{j-1}^T d_{j-1}} \right) d_{j-1}. \text{ (If } j = 1, \text{ set } d_1 = -g_1.)$$

$$x_{j+1/2} = x_j + \alpha_j d_j \text{ (typically, } \alpha_j = 1). \text{ Evaluate } g_{j+1/2}.$$

$$\alpha_j^* = -\frac{g_j^T(x_{j+1/2} - x_j)}{d_j^T(g_{j+1/2} - g_j)}$$

$$x_{j+1} = x_j + \alpha_j^* d_j. \text{ Evaluate } g_{j+1}.$$

end

4.3 Properties

The main properties of the conjugate gradient method, applied to strictly convex quadratics and using exact line searches, are *inherent* in the foregoing inductive development and require no further formal proof. They can be summarized as follows:

1. Prior to termination, each search direction d_j is a direction of descent and each step length α_j^* is a positive number. Each iterate x_{j+1} lies in the affine space $\{x_1 + \text{span}\{d_1, \ldots, d_j\}\}, j \leq n$.

2. Search direction vectors are conjugate, i.e., $d_i^T A d_j = 0 \ \forall \ i \neq j$.

3. Each gradient vector is orthogonal to all previous search directions, i.e., $g_{j+1}^T d_i = 0, 1 \leq i \leq j, 1 \leq j \leq n$. Thus x_{j+1} is the minimizing point in the affine space $\{x_1 + \text{span}\{d_1, \ldots, d_j\}\}, j \leq n$.

4. Gradient vectors at distinct iterates are orthogonal, i.e., $g_i^T g_j = 0 \ \forall \ i \neq j$.

9

5. $\operatorname{span}\{d_1, \ldots, d_j\} = \operatorname{span}\{g_1, \ldots, g_j\} = \operatorname{span}\{g_1, Ag_1, \ldots, A^{j-1}g_1\}$, $1 \leq j \leq n$. The last equality follows from repeated use of expression (3) along with $d_1 \parallel g_1$.

6. Finite termination at the optimal solution x^* occurs in, *at most, n* steps. This result can be further strengthened as follows: Suppose A has m *distinct* eigenvalues. Then the minimum polynomial of A has degree m, and $j = m$ is the least index for which the Krylov sequence of vectors $\{g_1, Ag_1, \ldots, A^j g_1\}$ are linearly dependent. It follows from properties 3 and 5 above that g_{m+1} is zero. Thus, the CG method converges in exactly m steps.[1]

4.4 Alternative Expressions for the CG Direction

Since line searches are assumed to be exact and gradient vectors at different iterates have been shown to be orthogonal, the numerator and denominator of the quantity

$$r_{j-1,j} = \frac{y_{j-1}^T g_j}{y_{j-1}^T d_{j-1}}$$

in expression (35) can be reexpressed as follows:

$$
\begin{aligned}
y_{j-1}^T g_j &= (g_j - g_{j-1})^T g_j \\
&= g_j^T g_j,
\end{aligned}
\tag{36}
$$

and

$$
\begin{aligned}
y_{j-1}^T d_{j-1} &= (g_j - g_{j-1})^T d_{j-1} \\
&= -g_{j-1}^T d_{j-1} \\
&= -g_{j-1}^T (-g_{j-1} + r_{j-2,j-1} d_{j-2}) \\
&= g_{j-1}^T g_{j-1}
\end{aligned}
\tag{37}
$$

Different combinations of (36) and (37) give alternative expressions for (35), and the three most common are as follows:

- *Hestenes-Stiefel*: $r_{j-1,j} = y_{j-1}^T g_j / y_{j-1}^T d_{j-1}$.

- *Fletcher-Reeves*: $r_{j-1,j} = g_j^T g_j / g_{j-1}^T g_{j-1}$.

- *Polak-Ribiere*: $r_{j-1,j} = y_{j-1}^T g_j / g_{j-1}^T g_{j-1}$.

In the present setting of convex quadratic functions and exact arithmetic, they are entirely equivalent, but they differ in a more general context as will be discussed in Chapter 4, Section 3.

---

[1] The proof is terse, and more detail on the terms used and this result are given by Wilkinson [1965] and Luenberger [1984], respectively.

5   The Quasi-Newton Relation

We now turn to the other key modern development of unconstrained nonlinear minimization, the quasi-Newton Method, in the setting of strictly convex quadratic functions. This method, which seeks to infer Hessian information from gradient information, will also highlight the discussion at the end of Section 2.1 on differences between the optimization and linear equation-solving contexts. This discussion will be concluded in Section 11.

Consider two distinct points, say x and x_+, obtained typically by a step-length procedure (Section 2.2) along some given direction d. Let g and g_+ be the corresponding gradients at these two points and, as usual, let $s = x_+ - x$ and $y = g_+ - g$ denote the step and corresponding gradient change. From (4),

$$As = y. \tag{38}$$

Any symmetric $n \times n$ matrix, say M_+, which shares this important property of the Hessian matrix, i.e.,

$$M_+s = y \tag{39}$$

is said to satisfy the *Quasi-Newton Relation* (or QN Relation) with respect to the pair (s, y). Note that, at the outset, we do *not* insist that M_+ be positive definite.

6   The Hereditary QN Property

Now, instead of a single pair (s, y), consider a set of pairs $\{(s_i, y_i) : 1 \le i \le j-1\}$ with $1 < j \le n + 1$, corresponding to nonzero steps s_i along $(j - 1)$ linearly independent directions d_i. These steps could be generated by a sequential step-length procedure as in Section 2.3, but initially we do not even require that they be contiguous. Obviously, $As_i = y_i, 1 \le i \le j - 1$. Now, suppose we are in possession of a symmetric matrix, say M_j, which satisfies the QN relation with respect to $(s_i, y_i), i \le i \le j - 1$, i.e.,

$$M_j s_i = y_i, \quad 1 \le i \le j - 1. \tag{40}$$

Thus,

$$(M_j - A)s_i = 0, \quad 1 \le i \le j - 1, \tag{41}$$

In particular, when there are n linearly independent steps, i.e., $(j-1) = n$ then $M_{n+1} = A > 0$ and M_{n+1} is thus uniquely determined and positive definite.

When $(j - 1) < n$, the equations (41) imply that $M_j s = As = y$ for any vector $s \in \text{span}\{s_1, \ldots, s_{j-1}\}$, so M_j satisfies the QN relation for *any* step s in $\text{span}\{s_1, \ldots, s_{j-1}\}$ with corresponding gradient change y. The larger the number $(j - 1)$, the "closer" is M_j to A in the sense of (41).

11

Consider now an additional pair, say, (s_j, y_j), such that $s_j \neq 0$ and s_j is *not* in the span$\{s_1, \ldots, s_{j-1}\}$. Usually, $M_j s_j \neq y_j$, but clearly $As_j = y_j$ continues to hold for a quadratic. For any i, $1 \leq i \leq j - 1$:

$$
\begin{aligned}
(y_j - M_j s_j)^T s_i &= y_j^T s_i - s_j^T M_j s_i \\
&= y_j^T s_i - s_j^T y_i \text{ using (40)} \\
&= s_j^T A s_i - s_j^T A s_i \text{ using } As_j = y_j \text{ and } As_i = y_i \\
&= 0.
\end{aligned}
$$

Thus the vector $(y_j - M_j s_j)$ is *orthogonal* to all previous steps $s_1, \ldots, s_{j-1}$, i.e.,

$$(y_j - M_j s_j)^T s_i = 0, \ 1 \leq i \leq j - 1. \tag{42}$$

This relation has important implications as we shall see in the next section.

We seek to revise or update the given matrix M_j so as to obtain a symmetric matrix M_{j+1} that satisfies

$$M_{j+1} s_j = y_j \tag{43}$$

and also *preserves* the QN relations (40) whenever possible, i.e. we want M_{j+1} to satisfy the QN relation with respect to the latest pair (s_j, y_j) and to possess the so-called *hereditary QN property* with respect to earlier pairs (s_i, y_i), namely, $M_j s_i = y_i$ implies $M_{j+1} s_i = y_i, 1 \leq i \leq j - 1$.

7   Low-Rank Updates to Satisfy the QN Relation

We will consider ways to update M_j by adding to it a symmetric matrix of low rank. The rationale for using a matrix of low rank is that it represents a "minimal" change to M_j, one that is simple and computationally inexpensive to implement.

7.1   The Symmetric Rank-1 (SR1) Update

An arbitrary symmetric rank-1 matrix must be of the form γuu^T, where $u \in \Re^n$ is an arbitrary vector and γ is a real number. For later comparison with rank two updates, note that γuu^T can be expressed as $u[\gamma]u^T$, where $[\gamma]$ represents an arbitrary 1×1 real matrix.

Suppose M_j is symmetric and satisfies $M_j s_j \neq y_j$. Consider an additive modification or *update* of M_j of the form

$$M_{j+1} = M_j + \gamma uu^T. \tag{44}$$

Since we want M_{j+1} to satisfy $M_{j+1} s_j = y_j$, expression (44) implies that

$$M_j s_j + \gamma(u^T s_j)u = y_j.$$

12

Thus u must be parallel to $y_j - M_j s_j$. Absorb the scalar that makes these two vectors equal into γ, denote this new quantity by γ_j, and reexpress (44) as

$$M_{j+1} = M_j + \gamma_j(y_j - M_j s_j)(y_j - M_j s_j)^T.$$

The unknown γ_j is determined by $M_{j+1} s_j = y_j$, i.e.,

$$\gamma_j = \frac{1}{(y_j - M_j s_j)^T s_j}, \tag{45}$$

and it is finite whenever $(y_j - M_j s_j)^T s_j \neq 0$.

Thus the SR1 update[2] is

$$M_{j+1} = M_j + \frac{(y_j - M_j s_j)(y_j - M_j s_j)^T}{(y_j - M_j s_j)^T s_j} \tag{46}$$

and it exists whenever $(y_j - M_j s_j)^T s_j \neq 0$. It is clear from the foregoing derivation that the SR1 update is *unique*, i.e. it is the only symmetric rank-one augmentation of M_j that satisfies the QN relation with respect to the pair (s_j, y_j).

Observe that an immediate consequence of the relation (42) is that the SR1 update possesses the hereditary QN property on a convex quadratic function, i.e., assuming that the matrix M_j already satisfies the relations $M_j s_i = y_i, 1 \leq i \leq j-1$ then (46) and (42) imply that $M_{j+1} s_i = y_i, 1 \leq i \leq j-1$. By construction, $M_{j+1} s_j = y_j$. Thus $M_{j+1} s_i = y_i, 1 \leq i \leq j$.

A simple mathematical procedure based on the SR1 update that seeks to infer the Hessian matrix A of f is as follows:

Procedure SR1:

Given $M_1 = I$ and a set of n linearly independent steps s_j with corresponding gradient change $y_j, 1 \leq j \leq n$:

for $j = 1, \ldots, n$

    if $(y_j - M_j s_j)^T s_j \neq 0$ *then*

        $M_{j+1} = M_j + \frac{(y_j - M_j s_j)(y_j - M_j s_j)^T}{(y_j - M_j s_j)^T s_j}$

    else

        $M_{j+1} = M_j.$

    endif

end

If the update exists at each iteration of this procedure, as will usually be the case, i.e. if $(y_j - M_j s_j)^T s_j \neq 0, 1 \leq j \leq n$ then $M_{n+1} = A > 0$ and x^*

---

[2] For later reference, note that the derivation of the update does *not* use properties of a quadratic function, i.e., it holds in general.

can be found from $x^* = x_{n+1} - M_{n+1}^{-1} g_{n+1}$. However, the procedure SR1 is not fail-safe.

Note also that intermediate updates $M_j, 1 \le j \le n$ need not be positive definite, and in general, $M_j > 0$ does not imply that $M_{j+1} > 0$, i.e., the SR1 update does *not* possess the *hereditary positive definiteness property*.[3]

Consequently, a search direction d_j obtained by solving $M_j d_j = -g_j$, when it exists, is *not* necessarily a direction of descent.

Finally assume that M_j and its update M_{j+1} are both nonsingular and define

$$W_j = M_j^{-1}, \quad W_{j+1} = M_{j+1}^{-1}.$$

Then W_{j+1} can be obtained by directly updating W_j, instead of inverting M_{j+1}, as follows:

$$W_{j+1} = W_j + \frac{(s_j - W_j y_j)(s_j - W_j y_j)^T}{(s_j - W_j y_j)^T y_j}. \tag{47}$$

This can be verified by multiplying W_{j+1} and M_{j+1} directly or by using the well-known Sherman-Morrison formula, which is defined in Chapter 2, Section 5. It is an axiom of numerical linear algebra that one avoids *computing* the inverse of a matrix whenever possible, but (47) is useful for mathematical purposes.

7.2   The Basic or B-Update

The SR1 update is unique. Thus, in order to broaden the approach, let us consider additively augmenting M_j by a symmetric matrix of rank two, in order to obtain greater generality and, in particular, the property of hereditary positive definiteness. This update takes the general form

$$M_{j+1} = M_j + [u, v] \Lambda [u, v]^T, \tag{48}$$

where $u, v \in \Re^n$ are arbitrary vectors and

$$\Lambda = \begin{bmatrix} \gamma_{11} & \gamma_{12} \\ \gamma_{21} & \gamma_{22} \end{bmatrix}$$

is an arbitrary 2×2 symmetric matrix with $\gamma_{12} = \gamma_{21}$. Henceforth, in this subsection, we shall also assume that $M_j > 0$, i.e., M_j is positive definite, and $M_j s_j \ne y_j$.

In order to satisfy the QN relation, $M_{j+1} s_j = y_j$, we must have

$$M_j s_j + [u, v] \Lambda \begin{bmatrix} u^T s_j \\ v^T s_j \end{bmatrix} = y_j. \tag{49}$$

This immediately suggests the choice $u = M_j s_j$ and $v = y_j$, i.e., the update:

$$M_{j+1} = M_j + [M_j s_j, y_j] \Lambda [M_j s_j, y_j]^T. \tag{50}$$

---

[3] Note that $M_j > 0$ and $(y_j - M_j s_j)^T s_j > 0$ does ensure that $M_{j+1} > 0$, because, under these assumptions, $z^T M_j z > 0$ implies $z^T M_{j+1} z > 0 \ \forall \ z \in \Re^n, z \ne 0$.

To further simplify the construction, it would be quite natural to restrict Λ to the set of *diagonal* matrices, i.e., take $\gamma_{12} = \gamma_{21} = 0$. Then from (49),

$$M_j s_j + \gamma_{11}(s_j^T M_j s_j) M_j s_j + \gamma_{22}(y_j^T s_j) y_j = y_j. \tag{51}$$

Whenever $M_j s_j$ and y_j are linearly independent, expression (51) then requires the choice $\gamma_{11} = -1/(s_j^T M_j s_j)$ and $\gamma_{22} = 1/(y_j^T s_j)$. Since $M_j > 0$ and $A > 0$ and $s_j \neq 0$, we have $s_j^T M_j s_j > 0$ and $y_j^T s_j = s_j^T A s_j > 0$.

This defines the *Basic* or B-Update[4], namely,

$$M_{j+1} = M_j + [M_j s_j, y_j] \begin{bmatrix} -\frac{1}{s_j^T M_j s_j} & 0 \\ 0 & \frac{1}{y_j^T s_j} \end{bmatrix} [M_j s_j, y_j]^T, \tag{52}$$

or equivalently,

$$M_{j+1} = M_j - \frac{(M_j s_j)(M_j s_j)^T}{s_j^T M_j s_j} + \frac{y_j y_j^T}{y_j^T s_j}. \tag{53}$$

When $M_j s_j$ and y_j are linearly dependent then the B-Update must reduce to the SR1 update, because the latter is unique. This can be verified directly. In order to obtain the SR1 update when $M_j s_j$ and y_j are linearly independent requires other choices for the parameters defining Λ in (48) or (50). There are many other possible choices and we shall return to this topic in Chapters 4 and 6. The B-Update has proved to be very satisfactory in practice. Its derivation here has been ad hoc, but in Chapter 2, Section 4.2 we shall see that it can be approached in a more general and, at the same time, a conceptually more satisfying way.

The B-update possesses the hereditary positive definiteness property, namely, $M_j > 0$ implies $M_{j+1} > 0$. This can be seen as follows.[5]

Let $M_j > 0$ and take any $z \in \Re^n$, $z \neq 0$. Then

$$\begin{aligned} z^T M_{j+1} z &= z^T M_j z - \frac{(z^T M_j s_j)^2}{s_j^T M_j s_j} + \frac{(z^T y_j)^2}{y_j^T s_j} \\ &= \frac{(z^T M_j z)(s_j^T M_j s_j) - (z^T M_j s_j)^2}{s_j^T M_j s_j} + \frac{(z^T y_j)^2}{y_j^T s_j}. \end{aligned}$$

Since $M_j > 0$, this matrix has a Cholesky factorization $M_j = R_j^T R_j$, where R_j is a nonsingular upper triangular matrix. Thus

$$z^T M_{j+1} z = \frac{\|R_j z\|_2^2 \|R_j s_j\|_2^2 - ((R_j s_j)^T (R_j z))^2}{s_j^T M_j s_j} + \frac{(z^T y_j)^2}{y_j^T s_j}. \tag{54}$$

---

[4] For later reference, note that we only use properties of a quadratic in the derivation to assert that $y_j^T s_j > 0$. Thus, whenever the pair (s_j, y_j) satisfies this condition, as can always be assured by a line search, then the foregoing derivation of the B-Update *holds in general*.

[5] When the pair (s_j, y_j) satisfies the condition $y_j^T s_j > 0$ then the proof of hereditary positive definiteness holds for arbitrary smooth functions.

The first term is nonnegative using the Cauchy-Schwartz inequality and $s_j^T M_j s_j > 0$. If $R_j s_j$ is parallel to $R_j z$, i.e., $s_j = \mu z$, $\mu \neq 0$ so the first term of (54) is zero, then the second term of this expression is $\mu^2 (y_j^T s_j) > 0$. Thus, in all cases, $z^T M_{j+1} z > 0$. Therefore, $M_j > 0$ implies that $M_{j+1} > 0$. $\square$.

However, the B-update does *not* possess the hereditary QN property, without further assumptions. Suppose that $M_j s_i = y_i, 1 \leq i \leq j-1$. Then (42) implies that $(y_j - M_j s_j)$ is orthogonal to $s_1, \ldots, s_{j-1}$. *Let us additionally require that the step s_j be conjugate to $s_1, \ldots, s_{j-1}$* , i.e., that y_j be orthogonal to $s_1, \ldots, s_{j-1}$. Then $M_j s_j$ must also be orthogonal to $s_1, \ldots, s_{j-1}$. It then follows immediately from (53) that $M_{j+1} s_i = y_i, 1 \leq i \leq j-1$, i.e., M_{j+1} has the hereditary QN property.[6] $\square$.

A simple finite procedure for inferring the Hessian matrix A of f using the B-update is as follows:

Procedure B:

Given $M_1 = I$ and n mutually *conjugate* steps $s_j \neq 0$ with corresponding y_j, $1 \leq j \leq n$:

for $j = 1, \ldots, n$

$$M_{j+1} = M_j - \frac{(M_j s_j)(M_j s_j)^T}{s_j^T M_j s_j} + \frac{y_j y_j^T}{y_j^T s_j}$$

end

In contrast to Procedure SR1, the foregoing Procedure B cannot break down in exact arithmetic and, after n steps, $M_{n+1} = A$ and $x^* = x_{n+1} - M_{n+1}^{-1} g_{n+1}$. Also $M_1 = I > 0$ and thus the hereditary positive definiteness property implies that $M_j > 0, 1 \leq j \leq n+1$.

Finally, in expression (53), define $W_j = M_j^{-1}$ and $W_{j+1} = M_{j+1}^{-1}$. One can obtain W_{j+1} by updating W_j directly, rather than inverting M_{j+1}, as follows:[7]

$$W_{j+1} = \left(I - \frac{s_j y_j^T}{y_j^T s_j} \right) W_j \left(I - \frac{s_j y_j^T}{y_j^T s_j} \right)^T + \frac{s_j s_j^T}{y_j^T s_j}. \tag{55}$$

This expression can be verified by multiplying M_{j+1} and W_{j+1} directly or by using the Sherman-Morrison formula twice. (See Chapter 2, Section 5 for a definition of the latter.)

Suppose *Procedure B* is modified in the obvious way to develop successive approximations to the inverse Hessian matrix A^{-1} based on expression (55),

---

[6] Observe that in constrast to the hereditary positive definiteness property, the hereditary QN property generally applies only to a quadratic function. See also Chapter 3, Section 3.1.

[7] Identical comments to those at the end of Subsection 7.1, regarding use of the inverse for mathematical rather than computational purposes, apply here as well.

where $W_1 = I$. Then for any j, $1 \leq j \leq n$,

$$W_{j+1} = \left(I - \sum_{i=1}^{j} \frac{s_i y_i^T}{y_i^T s_i}\right) W_1 \left(I - \sum_{i=1}^{j} \frac{s_i y_i^T}{y_i^T s_i}\right)^T + \sum_{i=1}^{j} \frac{s_i s_i^T}{y_i^T s_i}. \tag{56}$$

The proof of this useful mathematical expression is by induction as follows:
Suppose after $(j-1)$ updates, the matrix W_j is given by

$$W_j = \left(I - \sum_{i=1}^{j-1} \frac{s_i y_i^T}{y_i^T s_i}\right) W_1 \left(I - \sum_{i=1}^{j-1} \frac{s_i y_i^T}{y_i^T s_i}\right)^T + \sum_{i=1}^{j-1} \frac{s_i s_i^T}{y_i^T s_i}. \tag{57}$$

This is the induction hypothesis, and, for convenience, define $b_i = y_i^T s_i$ and $P_{j-1} = \left(I - \sum_{i=1}^{j-1} \frac{s_i y_i^T}{b_i}\right)$. Thus

$$W_j = P_{j-1} W_1 P_{j-1}^T + \sum_{i=1}^{j-1} \frac{s_i s_i^T}{b_i}.$$

Then, using expression (55), perform an additional update:

$$\begin{aligned}
W_{j+1} &= \left(I - \frac{s_j y_j^T}{b_j}\right) P_{j-1} W_1 P_{j-1}^T \left(I - \frac{s_j y_j^T}{b_j}\right)^T \\
&\quad + \left(I - \frac{s_j y_j^T}{b_j}\right) \left(\sum_{i=1}^{j-1} \frac{s_i s_i^T}{b_i}\right) \left(I - \frac{s_j y_j^T}{b_j}\right)^T + \frac{s_j s_j^T}{b_j}.
\end{aligned}$$

Using mutual conjugacy of the steps, i.e., $y_j^T s_i = 0$, $i \neq j$, the foregoing expression simplifies to (56).

The induction hypothesis is obviously true for $j = 1$, which completes the proof. $\square$

Expression (56) demonstrates, for example, that the matrix W_{j+1} is invariant with respect to permutations of the steps $s_1, \ldots, s_j$. Thus, if the steps $s_1, \ldots s_j$ had been used in a *different order* in *Procedure B*, the same matrix W_{j+1} would have resulted after j updates. This result is obviously true for $j = 1$ and $j = n$, and we now see that it also holds for all intermediate values of j. We shall appeal to expression (56) again at the end of Sections 8 and 9.

8 The Quasi-Newton Method

In order to derive a method, we must now be more specific about the way that the conjugate steps and successive approximations to the optimal solution in the foregoing section are defined. Let us do this in two stages.

17

First, instead of prespecified *steps* $s_j, 1 \le j \le n$, let us assume we are given a starting point x_1 and a set of conjugate *direction vectors*, $d_1, \ldots, d_n$, and let us now *interleave* iterations of the foregoing Procedure B with iterations of a sequential exact line search (Section 2.3). Thus at an iterate x_j, with M_j in hand, perform an exact line search along direction d_j to obtain α_j^*. (See (16) and recall that the step length can be of either sign.) Let $x_{j+1} = x_j + \alpha_j^* d_j$ and $s_j = x_{j+1} - x_j$. Then use s_j and the corresponding y_j to update M_j to M_{j+1}.

Second, drop the assumption that the direction vectors are prespecified, and instead, define the direction d_j used at iterate x_j as follows:

$$d_j = -M_j^{-1} g_j. \tag{58}$$

The rationale for this choice is that the Newton direction at x_j is $-A^{-1}g_j$ and M_j is the best available approximation to A. Relation (58) is called the **Quasi-Newton Direction**. (Note that d_j would be computed by solving the associated system of equations and *not* by inverting M_j. Expression (58) involving the inverse is a mathematical convenience.) If $M_j > 0$, which can be assured by using the B-update, the direction d_j is a *direction of descent* and the step length along it to the next iterate is positive.

We have now replaced *Procedure B* by the Quasi-Newton method in its most basic form. When stripped of detail concerning, for example, the detection of premature termination, it can be summarized as follows:

Procedure QN/B:

Set $M_1 = I$. Given x_1 and associated g_1:

for $j = 1, \ldots, n$

    Solve $M_j d_j = -g_j$ to obtain d_j.

    $x_{j+1/2} = x_j + \alpha_j d_j$ (typically, $\alpha_j = 1$). Evaluate $g_{j+1/2}$.

    $\alpha_j^* = -\dfrac{g_j^T(x_{j+1/2} - x_j)}{d_j^T(g_{j+1/2} - g_j)}$.

    $x_{j+1} = x_j + \alpha_j^* d_j$.

    Evaluate g_{j+1}; $s_j = x_{j+1} - x_j$ and $y_j = g_{j+1} - g_j$.

    $M_{j+1} = M_j - \dfrac{(M_j s_j)(M_j s_j)^T}{s_j^T M_j s_j} + \dfrac{y_j y_j^T}{y_j^T s_j}$.

end

We verify by induction that the directions developed by Procedure QN/B are mutually conjugate as follows:

Suppose that directions $d_1, \ldots, d_{j-1}$ are mutually conjugate for some index $j \le n$. Obviously $s_1, \ldots, s_{j-1}$ are then mutually conjugate. From Section 7.2, $M_j s_i = y_i$, $1 \le i \le j-1$ and $M_j > 0$. Suppose $g_j \ne 0$, else the procedure can be

18

terminated at $x_j = x^*$. Because line searches are exact, $g_j^T s_i = 0, 1 \leq i \leq j - 1$ (Section 3.1). Thus $\forall i$ such that $1 \leq i \leq j - 1$:

$$0 = g_j^T s_i = g_j^T (M_j^{-1} y_i) = (g_j^T M_j^{-1}) y_i = -d_j^T y_i.$$

Therefore d_j is conjugate to $s_1, \ldots, s_{j-1}$ and we have consequently verified that $d_1, \ldots, d_j$ are mutually conjugate. The result follows by induction. □

Thus all the desirable properties of Procedure B accrue to Procedure QN/B, in particular, the hereditary QN property and the hereditary positive-definiteness property. The procedure generates conjugate directions and terminates in, at most n, steps. In contrast to Procedure B, note that the foregoing Procedure QN/B can terminate prematurely; for example, when $A = I$, it terminates after a single step.

Finally, we verify that $\forall j$,

$$d_j \in \text{span}\{g_1, \ldots, g_j\}. \tag{59}$$

This result is intuitively obvious and in order to prove it formally, it is convenient to use the definition of d_j in the form $d_j = -W_j g_j$ where $W_j = M_j^{-1}$ and show the result inductively in tandem with the claim that each W_j can be expressed as follows:

$$W_j = I + \sum_{i \to 1}^{k[j]} p_i[j] q_i[j], \tag{60}$$

where

$$p_i[j] \in \text{span}\{g_1, \ldots, g_j\}, \ 1 \leq i \leq k[j], \tag{61}$$

$k[j]$ is an integer that depends on j, and the specific form of the vectors $p_i[j]$, $q_i[j] \in \Re^n \ \forall i, j$ is not of concern for the proof.

Assuming the results (59), (60) and (61) hold for some index j, let us prove them for index $j + 1$. From (55) along with this assumption, we directly obtain

$$W_{j+1} = I + \sum_{i=1}^{k[j+1]} p_i[j+1] q_i[j+1], \tag{62}$$

where

$$p_i[j+1] \in \text{span}\{g_1, \ldots, g_{j+1}\}, \ 1 \leq i \leq k[j+1], \tag{63}$$

for some integer $k[j+1]$. Then $d_{j+1} = -W_{j+1} g_{j+1}$ implies that

$$d_{j+1} \in \text{span}\{g_1, \ldots, g_{j+1}\}. \tag{64}$$

Thus the corresponding results hold for index $j + 1$ and the proof follows by induction. □

Note that the foregoing proof remains valid when line searches in Procedure QN/B are not exact. Under the assumption of exact line searches, which, in

19

turn, implies mutual conjugacy of the search direction vectors (established just after the statement of Procedure QN/B), the proof of (59) can be obtained much more simply by using the expression (57) with $W_1 = I$. It is straightforward to then verify that $d_j = -W_j g_j \in \text{span}\{g_1, \ldots, g_j\}$.

In Sections 7.2 and 8, we have concentrated on the B-Update, but there are other choices for the parameters in (50) that give quasi-Newton updates with similar properties. This will be taken up in Chapter 4, Section 2.1.3.

9   Relationship between Procedures CG and QN/B

Any method that is

- based on a sequential exact line search from a given starting point, say, x_1,

- develops search direction vectors d_j that are mutually conjugate,

- and restricts the choice of each d_j to lie in the span of the gradients at $x_1, \ldots, x_j$, i.e., $d_j \in \text{span}\{g_1, \ldots, g_j\}$

must generate search vectors at corresponding iterates that *can vary in length, but not in direction*, and hence must generate *identical* iterates. In other words, a method that has the foregoing three characteristics does not allow any essential freedom in the choice of search direction vectors (modulus their length). Procedures CG and QN/B are of this type. *Thus, in particular, procedure QN/B has the properties listed for procedure CG in Section 4.3.*

We now strengthen this observation by showing that the search direction vectors developed by the two procedures are *identical in both length and direction.*

Let d_j^B denote the directions developed by procedure QN/B, and let d_j^{CG} denote the corresponding directions developed by procedure CG. Since iterates are identical, we do not need to use separate symbols for gradients or steps in the two procedures. We know that d_j^B is parallel to the corresponding d_j^{CG} and we now show that

$$d_j^B = d_j^{CG}, 1 \leq j \leq n. \tag{65}$$

Again it is convenient to work with the inverse form of the update and as just noted, we can utilize all the properties given in Section 4.3. In particular,

$$g_k^T s_j = 0, \; k > j. \tag{66}$$

Let us assume for convenience that premature termination does not occur.

First we show by induction that

$$W_j g_k = g_k, \; j < k \leq n+1, \; 1 \leq j \leq n. \tag{67}$$

20

Assume the result is true for index j. Then the proof that it holds for index $(j + 1)$ follows by combining (55), (66), $g_i^T g_j = 0 \ \forall \ i \neq j$ and the hypothesis (67). Since $W_1 g_k = g_k, k > 1$, the result follows by induction.

Now return to the proof of the main assertion (65). By definition, $d_{j+1}^B = -W_{j+1} g_{j+1}$. Using (55) and the fact that line searches are exact,

$$
\begin{aligned}
d_{j+1}^B &= -W_j g_{j+1} + \frac{y_j^T W_j g_{j+1}}{y_j^T s_j} s_j \\
&= -W_j g_{j+1} + \frac{y_j^T W_j g_{j+1}}{y_j^T d_j^B} d_j^B \\
&= -g_{j+1} + \frac{y_j^T g_{j+1}}{y_j^T d_j^B} d_j^B \ \text{using (67)} \\
&= -g_{j+1} + \frac{y_j^T g_{j+1}}{y_j^T d_j^{CG}} d_j^{CG} \ \text{using } d_j^B \parallel d_j^{CG} \\
&= d_{j+1}^{CG}, \ 1 \leq j < n,
\end{aligned}
$$

and obviously $d_1^B = d_1^{CG}$. This is the desired result (65), which highlights the *structural correspondence* between the QN/B and CG methods. We shall see this again, in a more general setting, in Chapter 2. $\square$

The relationship can be strengthened by defining a more general version of Procedure QN/B, where the last operation of iteration j (M_j updated to M_{j+1}) is replaced as follows:

Let S_j denote the set consisting of the most recent step s_j along with a subset (possibly empty) of prior steps $s_1, \ldots, s_{j-1}$. Let $\kappa_j \geq 1$ be the number of steps in S_j. Set M_1 to the identity matrix I, and perform the B-Update in sequence over each step in S_j taken in any order. Set M_{j+1} (in the last line of Procedure QN/B) to the final matrix in the sequence, say M_{κ_j+1}.

Analogously to (56),

$$
W_{\kappa_j+1} = \left(I - \sum_{s_i \in S_j} \frac{s_i y_i^T}{y_i^T s_i} \right) I \left(I - \sum_{s_i \in S_j} \frac{s_i y_i^T}{y_i^T s_i} \right)^T + \sum_{s_i \in S_j} \frac{s_i s_i^T}{y_i^T s_i}, \tag{68}
$$

where $W_{\kappa_j+1} = M_{\kappa_j+1}^{-1}$. Using (68), it is straightforward to extend the foregoing proof to show that the search direction vectors developed by the QN/B procedure, modified as just described, are identical in length and direction to those developed by Procedure CG.

In the simplest case, suppose $S_j = \{s_j\}$, $\kappa_j = 1 \ \forall \ j$. Then

$$
M_{j+1} = M_2 = I - \frac{s_j s_j^T}{s_j^T s_j} + \frac{y_j y_j^T}{y_j^T s_j}
$$

and

$$W_{j+1} = W_2 = \left(I - \frac{s_j y_j^T}{y_j^T s_j}\right)\left(I - \frac{s_j y_j^T}{y_j^T s_j}\right)^T + \frac{s_j s_j^T}{y_j^T s_j}.$$

The resulting *memoryless* QN/B procedure is mathematically equivalent to the CG procedure. Of course, the numerical properties of the two procedures may be different (see also Section 11).

10 Preconditioned CG and QN/B Procedures

Consider a transformation of variables defined by

$$\tilde{x} = Px, \tag{69}$$

where P is any nonsingular matrix, often taken to be diagonal. Then the function (1) transforms to

$$\tilde{f}(\tilde{x}) = -b^T P^{-1}\tilde{x} + \frac{1}{2}\tilde{x}^T P^{-T} A P^{-1}\tilde{x} = -\tilde{b}^T\tilde{x} + \frac{1}{2}\tilde{x}^T\tilde{A}\tilde{x}, \tag{70}$$

where $\tilde{b} = P^{-T}b$ and $\tilde{A} = P^{-T}AP^{-1}$.

Analogously to Section 2.1, the gradient of the transformed quadratic at $\tilde{x}$ is given by

$$\tilde{g} = \nabla\tilde{f}(\tilde{x}) = -\tilde{b} + \tilde{A}\tilde{x} = P^{-T}(-b + AP^{-1}Px) = P^{-T}g, \tag{71}$$

and it follows directly from (69) and (71) (with an appropriate iteration subscript attached) that the quantities x_j, g_j, s_j, y_j and $g_{j+1}^T s_j$ transform to

$$\tilde{x}_j = Px_j, \quad \tilde{g}_j = P^{-T}g_j, \quad \tilde{s}_{j-1} = Ps_{j-1}, \quad \tilde{y}_{j-1} = P^{-T}y_{j-1}, \quad \tilde{g}_{j+1}^T\tilde{s}_j = g_{j+1}^T s_j. \tag{72}$$

Note also that directional derivatives or any other quantity defined by the inner product of a gradient vector and a direction vector (or point in the domain) are invariant under the transformation.

Suppose the conjugate gradient procedure of Section 4.2 is applied to the *transformed* quadratic (70). Then the relation defining the search direction vector is

$$\tilde{d}_j = -\tilde{g}_j + \left(\frac{\tilde{y}_{j-1}^T\tilde{g}_j}{\tilde{y}_{j-1}^T\tilde{d}_{j-1}}\right)\tilde{d}_{j-1}$$

This direction vector can then be transformed back and defined in terms of the original quantities using (72), to yield the *Preconditioned Conjugate Gradient* procedure as follows:

Procedure PCG:

Given x_1, associated gradient g_1 and preconditioner P:

for $j = 1, \ldots, n$

$$d_j = -(P^T P)^{-1} g_j + \left(\frac{y_{j-1}^T (P^T P)^{-1} g_j}{y_{j-1}^T d_{j-1}} \right) d_{j-1}. \ (\text{If } j = 1, \text{ set } d_1 = -(P^T P)^{-1} g_1.)$$

$x_{j+1/2} = x_j + \alpha_j d_j$ (typically, $\alpha_j = 1$). Evaluate $g_{j+1/2}$.

$$\alpha_j^* = -\frac{g_j^T (x_{j+1/2} - x_j)}{d_j^T (g_{j+1/2} - g_j)}.$$

$x_{j+1} = x_j + \alpha_j^* d_j$. Evaluate g_{j+1}.

end

Note that the quantity, say, $\hat{g}_j = (P^T P)^{-1} g_j$ used in the computation of the search direction d_j should be obtained by solving the linear system $(P^T P) \hat{g}_j = g_j$, and the transformation matrix P is often chosen to facilitate this operation as discussed at the end of this section.

In a similar vein, apply Procedure QN/B of Section 8 to the *transformed* quadratic (70). Then the B-Update in the last line of the procedure takes the form

$$\tilde{M}_{j+1} = \tilde{M}_j - \frac{(\tilde{M}_j \tilde{s}_j)(\tilde{M}_j \tilde{s}_j)^T}{\tilde{s}_j^T \tilde{M}_j \tilde{s}_j} + \frac{\tilde{y}_j \tilde{y}_j^T}{\tilde{y}_j^T \tilde{s}_j}.$$

Again transform back into the original quantities using (72), which taken along with the definition analogous to $A = P^T \tilde{A} P$, namely,

$$M_j = P^T \tilde{M}_j P \tag{73}$$

yields $M_1 = P^T \tilde{M}_1 P$, and reveals that the B-Update is *invariant*, i.e., after some simplification, it takes precisely the same form as before.

Thus we obtain the *Preconditioned* QN/B procedure as follows:

Procedure PQN/B:

Given x_1, associated gradient g_1 and nonsingular P, take $M_1 = P^T P$:

for $j = 1, \ldots, n$

Solve $M_j d_j = -g_j$ to obtain d_j.

$x_{j+1/2} = x_j + \alpha_j d_j$ (typically, $\alpha_j = 1$). Evaluate $g_{j+1/2}$.

$$\alpha_j^* = -\frac{g_j^T (x_{j+1/2} - x_j)}{d_j^T (g_{j+1/2} - g_j)}.$$

$x_{j+1} = x_j + \alpha_j^* d_j$

Evaluate g_{j+1}; $s_j = x_{j+1} - x_j$ and $y_j = g_{j+1} - g_j$.

$$M_{j+1} = M_j - \frac{(M_j s_j)(M_j s_j)^T}{s_j^T M_j s_j} + \frac{y_j y_j^T}{y_j^T s_j}.$$

end

Observe that apart from the the choice of initial Hessian approximation, all the other operations of Procedure QN/B of Section 8 are invariant.

Just as in Section 9, the Preconditioned CG and QN/B procedures, using the same choice for transformation matrix P, give search direction vectors that are identical in length and direction. Note also that all the properties of these procedures given in Section 4.3 are invariant except for items 4, 5 and 6, which are replaced by

4'. Gradient vectors at distinct iterates are orthogonal in the inner product defined by $(P^T P)^{-1}$, i.e., $g_i^T (P^T P)^{-1} g_j = 0$, $\forall i \neq j$.

5'. $\text{span}\{d_1, \ldots, d_j\} = \text{span}\{(P^T P)^{-1} g_1, \ldots, (P^T P)^{-1} g_j\}$, $1 \leq j \leq n$.

6'. Finite termination at the optimal solution x^* occurs in, *at most*, n steps. More specifically, finite termination occurs in as many steps as there are distinct eigenvalues of $\tilde{A} = P^{-T} A P^{-1}$.

The ideal choice of transformation matrix P is any factorization of the Hessian A of the form $A = P^T P$. For example, given the Cholesky factorization $A = LL^T$, where L is a nonsingular lower triangular matrix, one can take $P = L^T$. Then $\tilde{A} = I$ and the contours of the transformed quadratic become hyperspheres. In this case, the preconditioned preocedures using this choice of P terminate in one step along the Newton direction $-A^{-1} g_1$ at x_1. In general, P is chosen to reduce the condition number and cluster the eigenvalues of $\tilde{A}$ (as compared to A) without incurring excessive additional overhead in the operations of the algorithm; for example, P can be taken to be triangular, sparse or even diagonal (see also Chapter 6, Section 3).

11   Concluding Remarks

Our main objectives in this chapter have now been achieved, namely,

- to introduce and motivate the method of conjugate gradients and the quasi-Newton method (based on the SR1 and B-Updates) in the specialized setting of strictly convex quadratic minimization using exact line searches,

- to demonstrate the relationship between and the properties of the two methods in this specific context, and

- to provide a backdrop for the more general discussion of subsequent chapters.

This chapter serves to highlight correspondences and differences between the optimization problem and the computational linear algebra problem (solve the linear system $Ax^* = b, A > 0$). We can now complete the discussion begun at the end of Section 2.1.

First note that Procedure CG can be used as an iterative (as contrasted with direct or factorization-based) method for solving $Ax^* = b$. In this case the gradient g_j is the *residual* of the linear system $Ax_j - b$ at the iterate x_j. Since A is assumed to be available, the operation defining α_j^* can be replaced by $\alpha_j^* = -g_j^T d_j / d_j^T A d_j$. Different choices for mathematically equivalent expressions defining the search direction (see Section 4.4) and other operations in Procedure CG will have different stability and efficiency characteristics. Such issues and preconditioning strategies of the type mentioned in Section 10 are considered extensively in the computational linear algebra literature.

Turning now to the quasi-Newton method in the computational linear algebra context, it obviously does not make much sense to develop successive approximations M_j to A when the latter is on hand. Instead, one would view the QN method as a way of developing successive approximations W_j to the *inverse* of A. Alternatively, to avoid computation of an inverse, (a numerically undesirable approach), one can reorganize the QN method to update a *factorization* of M_j that converges in n steps to a factorization of A. The latter can then be used to efficiently solve the associated linear system.

Thus, along these lines, one could reformulate Procedure SR1 of Section 7.1 in the obvious way to replace computation of M_j by W_j. If the procedure does not fail, then $W_{n+1} = A^{-1}$ and $x^* = W_{n+1} b$. The procedure is much less efficient than other ways of developing the inverse of A and it is also numerically unstable. Thus it goes without saying that it has not enjoyed much use as a technique for solving a positive definite symmetric system of linear equations. A reformulation to update a factorization of M_j would be similarly doomed in the computational linear algebra context.

Likewise, Procedure QN/B can be reformulated to develop successive approximations W_j instead of M_j. In contrast to the SR1 procedure, the QN/B procedure modified in this way cannot fail, and it will find the solution of $Ax^* = b$, and, assuming premature termination does not occur, it will also find the inverse A^{-1} after n steps. Reformulations to update a Cholesky factorization of M_j will be referenced in Chapter 6 in the setting of nonlinear minimization. In the computational linear algebra context, they yield a QN method for developing a Cholesky factorization of A, but again there are more efficient linear equation solving techniques for achieving this purpose.

The foregoing remarks explain why the QN method has not enjoyed much use as a technique for solving linear equations, in contrast to the CG method. On the other hand, the *relationship* between the CG and QN methods could be useful in the computational linear algebra setting, because the CG method, reformulated as a *memoryless QN method*, may have better numerical properties than the original. This topic has not been extensively explored to date.

Apart from these few remarks, we will not pursue the computational linear algebra problem further in this monograph.

CHAPTER 2
THE METRIC-BASED CAUCHY PERSPECTIVE

1 Introduction

We now turn to the general unconstrained nonlinear minimization problem

$$\text{minimize}_{x \in \Re^n} \ f(x), \tag{1}$$

where $f : \Re^n \to \Re^1$ is a continuously differentiable function.

The classical method of Cauchy for solving (1) uses the steepest downhill direction at the current iterate. A suitable transformation of variables (equivalently: iterative preconditioning; reconditioning; change of metric) at each iteration, yields each of the other main classes of gradient-related methods of unconstrained nonlinear minimization. This is the unifying theme of the present chapter.

We follow an orderly progression of ideas that lead, in succession, to the Cauchy metric, Davidon's variable metric, the CG-metric and the Newton metric. The development is largely self-contained and sets the stage for the broader discussion of Chapters 4 and 6.

2 Steepest Descent

Given an approximation or current iterate, say $x \in \Re^n$, to a local minimum x^* of f with corresponding gradient vector $g = \nabla f(x) \neq 0$, consider the following problem:

$$\text{minimize}_{p \in \Re^n} \ \frac{g^T p}{||p||_2}. \tag{2}$$

This seeks the direction p, normalized to be of unit Euclidean length, along which the directional derivative is least, i.e. most negative. This is the direction along which the function locally decreases most rapidly. We can reexpress (2) as follows:

$$\text{minimize} \ g^T p$$
$$\text{s.t.} \ ||p||_2 = 1. \tag{3}$$

It is geometrically obvious that the unique solution of (2) or (3) is

$$p = -\frac{g}{||g||_2}, \tag{4}$$

and the standard Lagrangian optimality conditions can be used to formally establish this result. Thus the negative gradient is the steepest downhill direction

$d = -g$ at x and it is called the *Cauchy direction* or the direction of *steepest descent*.

The restriction of f to the direction $d = -g$ is the function $\rho : \Re^1 \rightarrow \Re^1$, where

$$\rho(\alpha) = f(x + \alpha d), \quad \alpha \in \Re^1, \tag{5}$$

and since d is a direction of descent at x, the next iterate can be obtained from

$$x_+ = x + \alpha^* d, \tag{6}$$

where

$$\alpha^* = \operatorname{argmin}_{\alpha \geq 0} \rho(\alpha).$$

The function (5) may have several local minima. In a conceptual setting, let us simply take the first positive local minimum for α^*. This corresponds to the first local minimizing point of f along the direction $d = -g \neq 0$ and ensures that $f(x_+) < f(x)$. We assume that the reader is familiar with *unidimensional* algorithms for minimizing a function like $\rho(\alpha)$, which are discussed in most elementary numerical analysis introductory textbooks (see the preface of this monograph for suitable references). The problem of solving (5) *in practice* is considered in Chapter 5, Section 2.1.

Let $g_+ = \nabla f(x_+) = \nabla f(x + \alpha^* d)$. Since $\rho(\alpha) = \nabla f(x + \alpha d)^T d$ and $\rho'(\alpha^*) = 0$, this implies that

$$g_+^T d = 0. \tag{7}$$

The procedure can now be repeated by making x_+ the current iterate. Thus $d_+ = -g_+$ and, from (7), $d_+^T d = 0$, i.e., successive search vectors are orthogonal. Since the function decreases monotonically along directions of steepest descent, it is plausible that the process will converge to a local minimizing point x^*. This must be established formally, an issue considered further in Chapter 5.

3 The Cauchy Metric

The Cauchy direction is ideal when the contours of the function f are hyperspheres. For example, consider the quadratic function $f(x) = -b^T x + \frac{1}{2} x^T x$, whose optimal solution is $x^* = b$. From *any* point x, a *unit* step along the negative gradient direction $d = -g = b - x$ would yield the optimal point $x^* = x + d = b$. For a general quadratic, the idea of *preconditioning* so as to make the transformed contours closer to hyperspheres was introduced in Chapter 1, Section 10. When the function f is nonquadratic, a similar idea carries over, i.e., precondition or transform the variables, in order to enhance the efficacy of the Cauchy direction in the transformed space.

We will first consider the transformation of variables

$$\tilde{x} = Dx, \tag{8}$$

where D is a given $n \times n$ diagonal matrix with positive elements $D_{ii}, i \leq i \leq n$ on the diagonal. Thus

$$\tilde{x}_i = D_{ii} x_i, \quad D_{ii} > 0, \quad 1 \leq i \leq n, \tag{9}$$

and obviously any direction vector d in the original variables transforms to

$$\tilde{d} = Dd. \tag{10}$$

Let $\tilde{f}(\tilde{x})$ denote the transformed function and $\tilde{g} = \nabla \tilde{f}(\tilde{x})$ denote the transformed gradient at $\tilde{x}$. Then for any i, $1 \leq i \leq n$,

$$(\nabla f)_i = \frac{\partial f}{\partial x_i} = \frac{\partial \tilde{f}}{\partial \tilde{x}_i} \frac{d\tilde{x}_i}{dx_i} = D_{ii} \frac{\partial \tilde{f}}{\partial \tilde{x}_i}$$

Thus

$$\tilde{g} = D^{-1} g. \tag{11}$$

In the transformed space, the steepest-descent direction is

$$\tilde{d} = -\tilde{g} = -D^{-1} g. \tag{12}$$

In the original variables, the descent direction d that corresponds to $\tilde{d}$ is obtained from (10),

$$Dd = -D^{-1} g, \quad d = -D^{-2} g. \tag{13}$$

Since $g^T d = -g^T D^{-2} g < 0$, the direction d is a direction of descent. The next iterate x_+ can then be obtained from (5) and (6) where d is defined by (13). Again, let $g_+ = \nabla f(x_+)$.

Since $\tilde{d} = -\tilde{g}$ is the steepest-descent direction in the transformed space, we know, in complete analogy to (2) and (4), that it can be obtained from the following problem and its solution:

$$\text{minimize}_{\tilde{p} \in \mathbb{R}^n} \frac{\tilde{g}^T \tilde{p}}{\|\tilde{p}\|_2}, \quad \tilde{p} = -\frac{\tilde{g}}{\|\tilde{g}\|_2}. \tag{14}$$

Since $\tilde{p} = Dp$ and $\tilde{g} = D^{-1} g$, this problem-solution pair can be expressed in the original variables as

$$\text{minimize}_{p \in \mathbb{R}^n} \frac{g^T p}{\|p\|_{D^2}}, \quad , \quad p = -\frac{D^{-2} g}{\|g\|_{D^{-2}}} = \frac{d}{\|g\|_{D^{-2}}}, \tag{15}$$

where $\|p\|_{D^2} = \sqrt{p^T D^2 p}$, $\|g\|_{D^{-2}} = \sqrt{g^T D^{-2} g}$ and d is given by (13). Thus we can interpret the direction d as the steepest-descent direction in the norm defined by the symmetric positive definite matrix D^2. *The associated metric will be called the Cauchy metric.* In the steepest-descent case we have $D = I$, corresponding to the Euclidean metric.

A simple mathematical procedure (Preconditioned Cauchy) for minimizing a smooth function f, stripped bare of detail, would take the following form:

Procedure PC:

Given a starting point x_1, gradient g_1 and fixed[1] diagonal preconditioning matrix $D > 0$:

for $j = 1, 2, \ldots$.

Solve $D^2 d_j = -g_j$ for d_j.

$\alpha_j^* = \mathrm{argmin}_{\alpha \geq 0} f(x_j + \alpha d_j)$.

$x_{j+1} = x_j + \alpha_j^* d_j$. Evaluate g_{j+1}.

end

4   The Davidon Variable Metric

Now consider a more general transformation of variables

$$\tilde{x} = Rx, \tag{16}$$

where R is any $n \times n$ nonsingular, and not necessarily symmetric, matrix. Then

$$\partial \tilde{x}_i / \partial x_j = R_{ij}, \;\; 1 \leq i \leq n, 1 \leq j \leq n, \tag{17}$$

and obviously any direction vector d in the original variables transforms to

$$\tilde{d} = Rd. \tag{18}$$

Again let $\tilde{f}(\tilde{x})$ denote the transformed function and $\tilde{g} = \nabla \tilde{f}(\tilde{x})$ denote the transformed gradient at $\tilde{x}$. Then

$$\tilde{g} = R^{-T} g, \tag{19}$$

which can be easily verified by the chain rule as follows:
For any j, $1 \leq j \leq n$,

$$(\nabla f)_j = \frac{\partial f}{\partial x_j} = \sum_{i=1}^{n} \frac{\partial \tilde{f}}{\partial \tilde{x}_i} \frac{\partial \tilde{x}_i}{\partial x_j} = \sum_{i=1}^{n} R_{ij} \frac{\partial \tilde{f}}{\partial \tilde{x}_i} = \sum_{i=1}^{n} R_{ji}^T \frac{\partial \tilde{f}}{\partial \tilde{x}_i} = (R^T \nabla \tilde{f})_j.$$

Thus $g = R^T \tilde{g}$, implying (19). $\square$

---

[1] The possibility of varying the diagonal matrix D at each iteration will be briefly addressed in Chapter 4, Section 4.

Henceforth, we attach the symbol 'tilde' to transformed quantities, and whenever it is necessary to explicitly identify the matrix used to define the transformation, we write $\tilde{x}[R]$ or $\tilde{g}[R]$ in (16) or (19).

In the transformed space, the steepest-descent direction is

$$\tilde{d} = -\tilde{g} = -R^{-T}g. \tag{20}$$

In the original variables, the descent direction d that corresponds to $\tilde{d}$ is obtained from (18),

$$Rd = -R^{-T}g, \quad d = -(R^T R)^{-1}g. \tag{21}$$

Define the positive definite matrices

$$M = R^T R \quad \text{and} \quad W = M^{-1}, \tag{22}$$

so (21) can be written more conveniently as

$$d = -M^{-1}g \quad \text{or} \quad d = -Wg, \tag{23}$$

and note that $g^T d = -g^T M^{-1} g < 0$, i.e., d is a direction of descent.

Since $\tilde{d} = -\tilde{g}$ is the steepest-descent direction in the transformed space, we again know, in analogy to (2) and (4), that it can be obtained from the following problem and its solution:

$$\text{minimize}_{\tilde{p} \in \Re^n} \frac{\tilde{g}^T \tilde{p}}{\|\tilde{p}\|_2}, \quad \tilde{p} = -\frac{\tilde{g}}{\|\tilde{g}\|_2}. \tag{24}$$

Since $\tilde{p} = Rp$ and $\tilde{g} = R^{-T}g$, this problem-solution pair can be expressed in the original variables as

$$\text{minimize}_{p \in \Re^n} \frac{g^T p}{\|p\|_M}, \quad , \quad p = -\frac{Wg}{\|g\|_W} = \frac{d}{\|g\|_W}, \tag{25}$$

where $\|p\|_M = \sqrt{p^T M p}$, $\|g\|_W = \sqrt{g^T W g}$ and d is given by (23). Thus, as in Section 3, we can interpret the direction d as the steepest-descent direction in the norm (and associated metric) defined by the symmetric positive definite matrix $M = R^T R$.

The next iterate x_+ can then be obtained from (5) or (6) where d is now defined by (23). Again, let $g_+ = \nabla f(x_+)$, and define

$$s = x_+ - x \quad \text{and} \quad y = g_+ - g. \tag{26}$$

In Davidon's variable metric approach, the matrix R defining the transformation is revised or updated *at each iteration*, using the information s and y gathered at that iteration, i.e., the problem is *iteratively reconditioned* in a suitable manner to which we now turn. We use the term 'reconditioning' to emphasize the fact that the operation is done iteratively, and to highlight the distinction with a more static 'preconditioning'.

4.1 The Variable-Metric Relation

As noted at the beginning of Section 3, the 'ideal' reconditioner, if it were to exist, would have the following property: given any two distinct reconditioned points of the space, a unit step from either point along the corresponding negative reconditioned gradient would lead to the *same* and, under these circumstances, to the optimal point x^* of f. Of course, this is not true of the reconditioner R at the particular points x and x_+, i.e., in general, $\tilde{x}[R] - \tilde{g}[R]$ and $\tilde{x}_+[R] - \tilde{g}_+[R]$ are different points. It is reasonable to require that the updated reconditioner R_+ have this property, at least at the two points x and x_+, i.e.,

$$\tilde{x}[R_+] - \tilde{g}[R_+] = \tilde{x}_+[R_+] - \tilde{g}_+[R_+],$$

or

$$\tilde{x}_+[R_+] - \tilde{x}[R_+] = \tilde{g}_+[R_+] - \tilde{g}[R_+]. \tag{27}$$

Thus, in the original variables,

$$R_+(x_+ - x) = R_+^{-T}(g_+ - g).$$

Using (26), this is

$$R_+ s = R_+^{-T} y, \tag{28}$$

and it can be equivalently expressed as

$$R_+^T R_+ s = y, \quad R_+ \text{ nonsingular}, \tag{29}$$

or as

$$M_+ s = y, \quad M_+ > 0, \tag{30}$$

where

$$M_+ = R_+^T R_+. \tag{31}$$

Since $s^T R_+^T R_+ s = s^T y$, a *necessary* condition for a nonsingular matrix R_+ satisfying (28) to exist is that

$$y^T s > 0. \tag{32}$$

Suppose the line search from x to x_+ is exact. Then $y^T s = (g_+ - g)^T s = -g^T s$, and because s is along a direction of descent, $-g^T s > 0$. Thus (32) is satisfied. In fact, (32) is a very weak condition that can be satisfied by a very inaccurate line search. All it requires is that the directional derivative at x_+ exceed that at x. See also Chapter 5, Section 2.1.

We shall see from the explicit construction of the next subsection that (32) is also a sufficient condition for the existence of R_+ satisfying (28). The condition (28) will be called the *Variable-Metric (VM) relation*. The first key idea of the variable metric approach is that the updated matrix R_+ *satisfy this VM relation*.

31

When more then one point is generated during a line search, it is important to observe that there are alternative choices for the quantity s and its corresponding y, for example, $(x_+ - x, \; g_+ - g)$ and $(x_+ - x(\alpha), \; g_+ - g(\alpha))$, where $x(\alpha)$ is some other point, typically between x and x_+, with associated gradient $g(\alpha)$. The steps $(x_+ - x)$ and $(x_+ - x(\alpha))$ are linearly dependent, but the corresponding gradient change vectors $(g_+ - g)$ and $(g_+ - g(\alpha))$ are not, unless the function is quadratic. In general, the updated matrix R_+ obtained using the first pair will differ from the matrix obtained using the second.

4.2 The Basic or B-Update

The second key idea in the variable metric approach is to revise R *using a matrix of low rank*. The basic or B-update makes the simplest possible augmentation of R by adding to it a matrix of rank one, i.e., a matrix of the general form uv^T, where u and v are arbitrary vectors in $\Re^n$. The updated matrix

$$R_+ = R + uv^T \tag{33}$$

is required to satisfy the variable metric relation (28) and is chosen to be as close as possible to R, measured by some appropriate matrix norm.

Thus, given s and y satisfying (32), we shall seek a solution $R_+ \in \Re^{n \times n}$ of the following *variational problem*:

$$\text{minimize } \|R_+ - R\| \tag{34}$$

s.t.

$$R_+ s = R_+^{-T} y, \tag{35}$$

where $\|.\|$ denotes a matrix norm.

Consider first the objective function (34). Using (33), it can be expressed as $\|uv^T\|$. Let us consider two natural choices for the matrix norm $\|.\|$.

1. *Frobenius Norm:* $\|A\|_F = \left[\sum_{i,j} a_{ij}^2 \right]^{1/2}$. Then

$$\|uv^T\|^2 \equiv \|uv^T\|_F^2 \;\; = \;\; \sum_{i=1}^{n}(u_1 v_i)^2 + \sum_{i=1}^{n}(u_2 v_i)^2 + \ldots + \sum_{i=1}^{n}(u_n v_i)^2$$

$$= \;\; \sum_{i=1}^{n}(v_i)^2 \, (u_1^2 + u_2^2 + \ldots + u_n^2)$$

$$= \;\; (u^T u)(v^T v).$$

2. *Spectral Norm:* $\|A\|_2 = \max_{z \in \Re^n} \left[\frac{\|Az\|_2}{\|z\|_2} \right]$. Then

$$\|uv^T\|^2 \equiv \|uv^T\|_2^2 \;\; = \;\; \max_{z \in \Re^n} \left[\frac{\|uv^T z\|_2}{\|z\|_2} \right]^2$$

$$
\begin{aligned}
&= (u^T u) \max_{z \in \Re^n} \frac{(v^T z)^2}{\|z\|_2^2} \\
&= (u^T u)(v^T v).
\end{aligned}
$$

In both cases, we obtain the same quantity. Henceforth, we shall replace the objective function (34) by $(u^T u)(v^T v)$. Note that the objective function is nonnegative and replacing it by its square does not alter the optimal solution of (34)-(35). Thus we consider the problem:

$$\text{minimize } (u^T u)(v^T v)$$

s.t.

$$(R + uv^T)s = (R + uv^T)^{-T} y. \tag{36}$$

This can be solved with the help of the standard Lagrangian optimality conditions. However, the special form of (36) also makes the problem amenable to solution by the following three-stage procedure that is especially tailored to it:

1. Rewrite (36) as

$$\text{minimize } (u^T u)(v^T v) \tag{37}$$

s.t.

$$(R + uv^T)^{-T} y = w \tag{38}$$

$$(R + uv^T)s = w \tag{39}$$

$$w^T w = y^T s > 0, \tag{40}$$

where the third constraint follows directly from the previous two.

2. Fix $w \neq 0$ at any value in $\Re^n$ and define the inner problem obtained from the objective function (37) and the reorganized first constraint (38), namely,

$$\text{minimize } (u^T u)(v^T v) \tag{41}$$

s.t.

$$(R + uv^T)^T w = y. \tag{42}$$

Denote the optimal solution of (41)-(42) by $u(w)$ and $v(w)$ and the corresponding optimal objective value by

$$[u(w)^T u(w)][v(w)^T v(w)].$$

Finding this optimal solution is easy. (42) implies that $v(u^T w) = y - R^T w$. Hence, assuming that $u^T w \neq 0$, we have $v = (y - R^T w)/(u^T w)$. Thus the *inner* problem can be written

$$\text{minimize}_{u \in \Re^n} \|y - R^T w\|_2^2 \frac{u^T u}{(u^T w)^2}.$$

33

In the optimal solution, $u(w)$ must be parallel to w, which also validates our assumption $u(w)^T w \neq 0$. Noting the form of the augmentation in (33), namely, uv^T, we may, without any loss of generality, take

$$u(w) = w \quad \text{which yields} \quad v(w) = \frac{y - R^T w}{w^T w}. \tag{43}$$

3. Restate the *outer* problem (37)-(40) as

$$\text{minimize}_{w \in \Re^n} [u(w)^T u(w)][v(w)^T v(w)] \tag{44}$$

s.t.

$$(R + u(w)v(w)^T)s = w \tag{45}$$

$$w^T w = y^T s > 0. \tag{46}$$

Fortunately, the constraints (45) - (46) do not allow much additional freedom of choice. The solution (43) and the constraint (45) imply that

$$\left[R + \frac{w}{w^T w}(y - R^T w)^T \right] s = w.$$

$$Rs + \frac{w}{w^T w}(y - R^T w)^T s = w.$$

Thus the existence of a feasible solution to (45) requires that w be parallel to Rs, i.e., $w = \tau Rs$ for some scalar τ. Using (46), $\|w\|_2 = \sqrt{y^T s}$, hence

$$\tau = \pm \frac{\sqrt{y^T s}}{\|Rs\|_2} = \pm \left(\frac{b}{a} \right)^{1/2},$$

where $a = s^T R^T Rs > 0$ and $b = y^T s > 0$.

Thus, optimizing (44)-(46) is trivial, because there are only two cases to consider, namely,

$$u(w) = \pm \left(\frac{b}{a} \right)^{1/2} Rs, \quad v(w) = \frac{y \mp \left(\frac{b}{a} \right)^{1/2} R^T Rs}{b}.$$

From (44), we must minimize $\|u(w)\|_2^2 \|v(w)\|_2^2$, which gives the solution

$$u(w) = \left(\frac{b}{a} \right)^{1/2} Rs, \quad v(w) = \frac{y - \left(\frac{b}{a} \right)^{1/2} R^T Rs}{b} \quad \text{when } y^T R^T Rs > 0 \tag{47}$$

and

$$u(w) = - \left(\frac{b}{a} \right)^{1/2} Rs, \quad v(w) = \frac{y + \left(\frac{b}{a} \right)^{1/2} R^T Rs}{b} \quad \text{when } y^T R^T Rs < 0. \tag{48}$$

34

Substituting these quantities into (33) gives the *Basic* or *B-Update*:

$$R_+ = R \pm \frac{Rs}{(ab)^{1/2}} \left(y \mp \left(\frac{b}{a} \right)^{1/2} R^T Rs \right)^T. \tag{49}$$

This completes the derivation.

It is worth noting that additional freedom in (33), for example, use of a rank-2 update, would make the outer problem analogous to (44)-(46) much less amenable to solution. In fact, when an augmentation of R of arbitrary rank is permitted, it can be shown that the optimal choice is a rank-2 update of R. □

Having revised our reconditioner in the way just described, the corresponding reconditioned steepest-descent direction is defined analogously to (23), i.e.,

$$d_+ = -(R_+^T R_+)^{-1} g_+ = -M_+^{-1} g_+ = -W_+ g_+, \tag{50}$$

where $M_+ = R_+^T R_+$ and $W_+ = M_+^{-1}$. A line search would be initiated at x_+ along d_+ and the updating procedure continued.

Since $R_+^T R_+$ determines the search direction, it is natural to form the matrix M_+ explicitly, i.e.,

$$M_+ = \left[R \pm \frac{Rs}{(ab)^{1/2}} \left(y \mp \left(\frac{b}{a} \right)^{1/2} R^T Rs \right)^T \right]^T \left[R \pm \frac{Rs}{(ab)^{1/2}} \left(y \mp \left(\frac{b}{a} \right)^{1/2} R^T Rs \right)^T \right]$$

where upper or lower signs are taken together. Simplifying, we obtain

$$M_+ = M - \frac{(Ms)(Ms)^T}{s^T Ms} + \frac{yy^T}{y^T s}, \tag{51}$$

Recall the definition $M = R^T R$ and note that the choice of signs makes no difference in M_+. *We discover that we have obtained the B-Update of Chapter 1, Section 7.2 after we equate M_j with M and M_{j+1} with M_+.* However, in contrast to that derivation, the present derivation is not ad-hoc. Another difference between the present setting and the quadratic setting of Chapter 1 is that the B-update does not normally have the hereditary QN property on general functions even when line searches are exact.[2]

Suppose that $y^T s > 0$ and R is nonsingular. The latter condition implies that $M > 0$. We showed in Chapter 1, Section 7.2 that $y^T s > 0$ and $M > 0$ imply $M_+ > 0$, a result that is independent of the quadratic setting. Finally, $M_+ > 0$ implies that R_+ is nonsingular. Thus we conclude that $y^T s > 0$ and R nonsingular imply that R_+ is nonsingular.

We can now state a conceptual variable-metric procedure based on the B-Update, stripped bare of unnecessary detail, as follows:

---

[2] Conditions under which the hereditary property can be achieved will be considered in Chapter 3.

35

Procedure VM/B:

Given a starting point x_1 and initial nonsingular reconditioner R_1:

for $j = 1, 2, \ldots$.

Solve $(R_j^T R_j) d_j = -g_j$ for d_j.

$\alpha_j^* = \text{argmin}_{\alpha \geq 0} f(x_j + \alpha d_j)$:

$x_{j+1} = x_j + \alpha_j^* d_j$. Evaluate g_{j+1}.

$s_j = x_{j+1} - x_j$; $y_j = g_{j+1} - g_j$; $a_j = s_j^T R_j^T R_j s_j$; $b_j = y_j^T s_j$.

$$R_{j+1} = R_j \pm \frac{R_j s_j}{(a_j b_j)^{1/2}} \left(y_j \mp \left(\frac{b_j}{a_j} \right)^{1/2} R_j^T R_j s_j \right)^T.$$

end

Since the foregoing is a descent procedure, it is plausible that it yields a point where the gradient of f vanishes. This is considered formally in Chapter 5.

5   The CG-Metric

Given a current iterate x with corresponding gradient g and a reconditioner R, a search direction d was defined by (22) and (23). A step s was taken along d to a new iterate x_+, with corresponding gradient g_+ and gradient change $y = g_+ - g$. Using R, s and y, the reconditioner was updated to R_+. This matrix replaced R, enabling the search procedure to be repeated at x_+.

Let us now consider the case when the reconditioner R *is discarded immediately after it has been used to define a search direction d*, so that the development of R_+ must start from scratch without prior information, i.e. it must be obtained by updating some simple matrix, for example, the identity matrix I. A plausible scenario for this case is when n is large and a matrix requiring $O(n^2)$ storage cannot be maintained. However, vectors that implicitly define a rank-2 update together with a simple matrix that it augments, for example, a diagonal matrix, can be stored, because this information only requires $O(n)$ storage. Thus, when $R = I$, (49) becomes

$$R_+ = I + \frac{s}{(ab)^{1/2}} \left(y - \left(\frac{b}{a} \right)^{1/2} s \right)^T \tag{52}$$

where $a = s^T s$, $b = y^T s$ and only the upper signs need be considered, i.e., expression (47), because $y^T R^T R s = y^T s > 0$.

The well-known *Sherman-Morrison formula* can be used to find the inverse of the foregoing matrix R_+. This result can be stated as follows:

Suppose $u, v \in \Re^n$ and the matrix $A \in \Re^{n \times n}$ is nonsingular. Then $A + uv^T$ is nonsingular if and only if

$$(1 + v^T A^{-1} u) \neq 0 \tag{53}$$

Furthermore,

$$(A + uv^T)^{-1} = A^{-1} - \frac{A^{-1} uv^T A^{-1}}{1 + v^T A^{-1} u}. \quad \square \tag{54}$$

Apply this lemma to (52). Then expression (53) becomes

$$1 + \frac{1}{(ab)^{1/2}} \left(y - \left(\frac{b}{a}\right)^{1/2} s \right)^T s = \left(\frac{b}{a}\right)^{1/2} \neq 0, \tag{55}$$

which, in turn, gives

$$R_+^{-1} = I - \frac{s}{b} \left(y - \left(\frac{b}{a}\right)^{1/2} s \right)^T. \tag{56}$$

The corresponding search direction d_+ is obtained by solving

$$(R_+^T R_+) d_+ = -g_+. \tag{57}$$

Since we do not want to form a matrix (or a triangular factor) explicitly as in Procedure VM/B, it is now convenient to use (56), i.e.,

$$d_+ = - \left[I - \frac{s}{b} \left(y - \left(\frac{b}{a}\right)^{1/2} s \right)^T \right] \left[I - \left(y - \left(\frac{b}{a}\right)^{1/2} s \right) \frac{s^T}{b} \right] g_+. \tag{58}$$

Note that the matrix R_+^{-1} in (56) is defined *implicitly* by the vectors s and y and the scalars a and b, and the direction d_+ is found using only dot products and scalar operations, i.e., *no matrix is explicitly formed*. Keeping this in mind, a simple limited-storage conceptual procedure based on (58) can be stated as follows:

Procedure LVM/B:

Given x_1 and corresponding g_1:

for $j = 1, 2, \ldots$.

$a_{j-1} = s_{j-1}^T s_{j-1}; \quad b_{j-1} = y_{j-1}^T s_{j-1}.$
(If $j = 1$ then a_{j-1} and b_{j-1} are not defined.)

$$d_j = -\left[I - \tfrac{s_{j-1}}{b_{j-1}}\left(y_{j-1} - \left(\tfrac{b_{j-1}}{a_{j-1}}\right)^{1/2} s_{j-1}\right)^T\right]\left[I - \left(y_{j-1} - \left(\tfrac{b_{j-1}}{a_{j-1}}\right)^{1/2} s_{j-1}\right)\tfrac{s_{j-1}^T}{b_{j-1}}\right]g_j.$$

(If $j = 1$, set $d_1 = -g_1$.)

$$\alpha_j^* = \operatorname{argmin}_{\alpha \geq 0} f(x_j + \alpha d_j).$$

$x_{j+1} = x_j + \alpha_j^* d_j$. Evaluate g_{j+1}.

$s_j = x_{j+1} - x_j; \quad y_j = g_{j+1} - g_j.$

end

Since line searches in the foregoing procedure are taken to be exact, we have $g_j^T s_{j-1} = 0$, and using this relation, the direction d_j simplifies considerably as follows:

$$d_j = -g_j + \frac{y_{j-1}^T g_j}{y_{j-1}^T s_{j-1}} s_{j-1} = -g_j + \frac{y_{j-1}^T g_j}{y_{j-1}^T d_{j-1}} d_{j-1}. \tag{59}$$

This is precisely the Hestenes-Stiefel conjugate gradient direction (1:35), but now it has been derived in a conceptually more general and satisfactory way. We may also observe that under the assumption of exact line searches, the Hestenes-Stiefel and Polak-Ribiere expressions for the CG direction (see Chapter 1, Section 4.4) are identical for *arbitrary smooth functions*, which can be seen as follows:

$$
\begin{aligned}
y_{j-1}^T d_{j-1} &= -g_{j-1}^T d_{j-1}, \text{ because } g_j^T d_{j-1} = 0 \\
&= -g_{j-1}^T\left(-g_{j-1} + \left(\frac{y_{j-2}^T g_{j-1}}{y_{j-2}^T d_{j-2}}\right) d_{j-2}\right) \\
&= g_{j-1}^T g_{j-1}, \text{ because } g_{j-1}^T d_{j-2} = 0.
\end{aligned}
$$

Thus (59) can be written in the equivalent Polak-Ribiere form

$$d_j = -g_j + \frac{y_{j-1}^T g_j}{g_{j-1}^T g_{j-1}} d_{j-1}. \quad \Box \tag{60}$$

Under the foregoing assumptions, the Fletcher-Reeves choice, which we may also note is always nonnegative, is not equivalent to the other two.

When line searches are *no longer required to be exact*, Procedure LVM/B can still be used, with α_j^* suitably replaced by a step-length routine that sufficiently reduces the function value, but the search direction d_j is no longer equivalent to (59) or (60). Indeed, the direction defined by (59) is then no longer guaranteed to be a direction of descent. To ensure descent, the point x_j and associated gradient g_j obtained by a line search along d_{j-1} must satisfy the condition $g_j^T d_j < 0$, i.e.,

$$\frac{y_{j-1}^T g_j}{y_{j-1}^T d_{j-1}} g_j^T d_{j-1} < g_j^T g_j. \tag{61}$$

Since $g_j^T d_{j-1} = 0$ when the line search is exact, the condition (61) can always be assured by making the line search sufficiently accurate.

In contrast, the direction defined by Procedure LVM/B is a direction of descent whenever $b_{j-1} = y_{j-1}^T s_{j-1} > 0$. This is a weaker condition on the line search along d_{j-1} than (61), and, of course, a necessary condition in a metric-based setting. The discussion on practicality will be resumed in Chapters 4 and 5.

Reverting to our simpler notation without iteration subscripts, observe that the direction defined by (57) is the direction of steepest descent in the metric defined by $R_+^T R_+$, where R_+ is given by (52). Thus, for obvious reasons, the reconditioner defined by (52) will be called the *Conjugate-Gradient Reconditioner* and the corresponding metric defined by $M_+ = R_+^T R_+$ will be called the *Conjugate-Gradient Metric*. From expression (51), the matrix $M_+ > 0$ is

$$M_+ = I - \frac{ss^T}{a} + \frac{yy^T}{b}. \tag{62}$$

where $a = s^T s > 0$ and $b = y^T s > 0$. As in Chapter 1, Section 7.2, $W_+ = M_+^{-1}$ is given by

$$W_+ = \left(I - \frac{sy^T}{b} \right) \left(I - \frac{sy^T}{b} \right)^T + \frac{ss^T}{b}. \tag{63}$$

6   The Newton Metric

Finally, let us consider the case when the Hessian matrix $H = \nabla^2 f(x)$ is available at the current iterate x. If H is indefinite, let $\bar{H}$ be a symmetric positive definite matrix obtained from H by modifying the latter in some "minimal' way. For example, if $\sigma_1 < 0$ is the smallest eigenvalue of H, then we can take $\bar{H} = H + \lambda I$, where $\lambda > -\sigma_1$. (There a numerous other ways to modify H as will be discussed in Chapter 4.) When $H > 0$, take $\bar{H} = H$.

Let $\bar{H} = \bar{R}^T \bar{R}$ be any factorization of $\bar{H}$. Obviously $\bar{R}$ is nonsingular and can be used as to define a reconditioner in a development completely analogous to that of Section 4. The corresponding steepest descent direction, in the metric defined by $\bar{H}$, is the solution of

$$\text{minimize}_{p \in \Re^n} \frac{g^T p}{\|p\|_{\bar{H}}} \tag{64}$$

and, analogously to (25), this direction is

$$p = -\frac{\bar{H}^{-1}g}{\|g\|_{\bar{H}^{-1}}},\qquad (65)$$

where $\|g\|_{\bar{H}^{-1}} = \sqrt{g^T \bar{H}^{-1} g}$.

The metric defined by $\bar{H} > 0$ will be called the *Newton Metric* and the search direction vector $d = -\bar{H}^{-1}g$ will be called the *Modified-Newton Direction*.

A conceptual *Procedure N* would be identical to Procedure PC of Section 3, with the fixed matrix D^2 replaced by the modified Hessian matrix at iteration j, say $\bar{H}_j$.

7 Notes

Section 3; The metric-based Cauchy perspective has its roots in the pioneering work of Davidon [1959].

Section 4.2: The derivation of the B-Update is a variant of an approach originated by Dennis and Schnabel [1981]. See also Zhang and Tewarson [1987] for more recent developments.

Section 5: The relationship between the BFGS (B-Update) and CG methods, on which many limited storage variants are founded, was developed by Nazareth [1976b] and extended by Buckley [1978].

CHAPTER 3
THE MODEL-BASED NEWTON PERSPECTIVE

1 Introduction

We again consider the general unconstrained minimization problem of Chapter 2, namely,

$$\text{minimize}_{x \in \Re^n} \ f(x), \tag{1}$$

where $f : \Re^n \to \Re^1$ is a smooth nonlinear function.

The classical method of Newton replaces the function at an approximate solution x by a quadratic function, which is derived, in turn, from a truncation of the Taylor expansion at x. Similarly, the model-based approach approximates the function f at the current approximate solution by a suitable unconstrained or constrained local appproximating model, which is then used to obtain an improving point. This is the unifying theme of the present chapter.

As with the preceeding two chapters, the development is largely self-contained and follows a logical progression of ideas that lead, in succession, to the Newton model, the quasi-Newton model, the CG-related model and the Cauchy model. The development complements that of Chapter 2, the reverse side of the coin so to speak, and completes the foundation for the broader discussion of the next three chapters.

2 The Newton Model

Given an approximate solution, say, $x \in \Re^n$, to a local minimum x^* of f, approximate the function f by $\psi : \Re^n \to \Re^1$, a quadratic function obtained from the non-constant terms, upto second order, of the Taylor expansion of f at x:

$$\psi(z) = g^T(z - x) + \frac{1}{2}(z - x)^T H(z - x), \tag{2}$$

where $g = \nabla f(x) \neq 0$ denotes the gradient vector and $H = \nabla^2 f(x)$ denotes the Hessian matrix or matrix of second partial derivatives $[\partial^2 f(x)/\partial x_i \partial x_j]$ at the current iterate x. Note that H is a symmetric matrix, which we now assume is available for use in a computational procedure.

When H is positive definite, the foregoing model ψ is strictly convex and has a well-defined minimum, namely,

$$z^* = x - H^{-1}g,$$

as in Chapter 1, Sec. 2.1. However, H can be indefinite, in which case ψ has no minimizing point and does not serve as a suitable model. An additional

41

constraint must be imposed to restrict the step length, leading to the following constrained local approximating model, termed *the Newton model*:

$$\text{minimize} \ \ g^T(z-x) + \frac{1}{2}(z-x)^T H(z-x) \tag{3}$$

s.t.

$$||z - x||_2 \leq \Delta, \tag{4}$$

where $\Delta = \Delta(x)$ denotes a positive number that can vary with the current iterate x and $||.||_2$ denotes the Euclidean norm used to define the ball[1] constraint. After analysing this model, we shall return to the choice of Δ.

2.1 Global Optimality Conditions

The point z^* is a global minimizing point of (3)-(4) if and only if there exists a number $\lambda^* \in \Re^1$, the *Lagrange multiplier* associated with the ball constraint, such that

$$(H + \lambda^* I)(z^* - x) \ = \ -g, \ \ \lambda^*(\Delta - ||z^* - x||_2) = 0 \ \ \text{with} \ \lambda^* \geq 0, \tag{5}$$
$$(H + \lambda^* I) \ \ \text{is} \ \ \text{positive semidefinite, and} \tag{6}$$
$$||z^* - x||_2 \ \leq \ \Delta. \tag{7}$$

This result can be established as follows:

First consider *necessity*. Suppose z^* is a global minimizing point for the model (3)-(4), which can be rewritten more conveniently using the definitions

$$w = z - x, \ \ w^* = z^* - x, \tag{8}$$

as follows:

$$\text{minimize} \ \ g^T w + \frac{1}{2} w^T H w \tag{9}$$

s.t.

$$||w||_2 \leq \Delta. \tag{10}$$

Then the standard first order optimality conditions (Karush-Kuhn-Tucker conditions) imply that

$$(H + \lambda^* I)w^* = -g, \ \lambda^*(\Delta - ||w^*||_2) = 0, \ \lambda^* \geq 0 \ \text{and} \ ||w^*||_2 \leq \Delta. \tag{11}$$

This is equivalent to (5) and (7) in the notation (8), so it remains to show (6).

When w^* is optimal, i.e., a minimizing point, and $||w^*||_2 < \Delta$ then the second condition of (11) implies that $\lambda^* = 0$. Thus, using the first condition of

---

[1] A more general ellipsoidal constraint of the form $||z - x||_D \leq \Delta$, where D is a diagonal scaling matrix with positive diagonal elements, can easily be substituted (see also Chapter 4, Section 1.2.2).

(11), $Hw^* = -g$, and w^* is an unconstrained minimizing point of the objective function (9). Then the standard second-order necessary condition for an *unconstrained* minimum implies that H must be positive semidefinite. Since $\lambda^* = 0$, it follows that $H + \lambda^* I$ is positive semidefinite, i.e., (6) holds.

Thus let us consider the case w^* optimal for the model and $||w^*||_2 = \Delta$. Then, for all points w on the boundary of (10), i.e., points such that $||w||_2 = \Delta = ||w^*||_2$,

$$g^T w + \frac{1}{2} w^T H w \geq g^T w^* + \frac{1}{2} w^{*T} H w^*$$
$$= g^T w^* + \frac{1}{2} w^{*T} H w^* + \frac{\lambda^*}{2}(w^{*T} w^* - w^T w).$$

From (11), $(H + \lambda^* I)w^* = -g$, and using this to replace g gives

$$-w^{*T}(H+\lambda^* I)w + \frac{1}{2} w^T H w \geq -w^{*T}(H+\lambda^* I)w^* + \frac{1}{2} w^{*T} H w^* + \frac{\lambda^*}{2}(w^{*T} w^* - w^T w).$$

Rearranging this expresssion gives

$$\frac{1}{2}(w - w^*)^T (H + \lambda^* I)(w - w^*) \geq 0 \ \forall \ w \ \text{s.t.} \ ||w||_2 = ||w^*||_2.$$

The closure of the set of directions $p = w - w^*$, for all w such that $||w||_2 = ||w^*||_2$, is the half-space $p^T w^* = ||w^*||_2^2$ and from this it follows that $(H + \lambda^* I)$ must be positive semidefinite. This establishes (6) and completes the proof of necessity of the conditions (5)-(7).

Next, to establish their *sufficiency*, assume that the conditions (5)-(7) hold. These are equivalent to (6) and (11) in the alternative notation (8). For any $p \in \Re^n$,

$$g^T(w^* + p) + \frac{1}{2}(w^* + p)^T(H + \lambda^* I)(w^* + p) = g^T w^* + \frac{1}{2} w^{*T}(H + \lambda^* I)w^* + g^T p$$
$$+ \frac{1}{2} p^T(H + \lambda^* I)p + w^{*T}(H + \lambda^* I)p.$$

Defining $w = w^* + p$ and using $(H + \lambda^* I)w^* = -g$, the above equation becomes

$$g^T w + \frac{1}{2} w^T(H + \lambda^* I)w = g^T w^* + \frac{1}{2} w^{*T}(H + \lambda^* I)w^* + \frac{1}{2} p^T(H + \lambda^* I)p.$$

Since $(H + \lambda^* I)$ is positive semidefinite,

$$g^T w + \frac{1}{2} w^T(H + \lambda^* I)w \geq g^T w^* + \frac{1}{2} w^{*T}(H + \lambda^* I)w^*.$$

Therefore,

$$g^T w + \frac{1}{2} w^T H w \geq g^T w^* + \frac{1}{2} w^{*T} H w^* + \frac{\lambda^*}{2}(w^{*T} w^* - w^T w).$$

43

From (11), $\lambda^* \|w^*\|_2^2 = \lambda^* \Delta^2$. Hence

$$g^T w + \frac{1}{2} w^T H w \geq g^T w^* + \frac{1}{2} w^{*T} H w^* + \frac{\lambda^*}{2}(\Delta^2 - w^T w).$$

Thus w^* minimizes the objective function (9) for all w such that $\|w\|_2 \leq \Delta$. Equivalently, z^* is a *global minimum* of the model (3) - (4). $\square$

2.2   Choice of Ball Radius

Now let us consider choices for Δ. When $H > 0$, the ball constraint (4) is not needed, because the objective function of the model is well defined. Thus a reasonable choice for Δ in this case, which, in effect, dispenses with the ball constraint, is

$$\Delta = \|H^{-1}g\|_2. \tag{12}$$

Note that Δ depends on H and g and is thus a function of the current iterate x.

For a general matrix H, it is natural to continue to define the size of the ball by (12), i.e., by the distance to the stationary point of the objective function, because the ball constraint is then rendered redundant when H happens to be positive definite. In the subsequent analysis, we shall consider the foregoing simple choice for Δ and return to other choices in subsequent chapters.

2.3   Extracting a Descent Direction

We now study the optimal z^*, λ^* as characterized by (5)-(7), the global optimality conditions of the Newton model, and we shall employ the following notation.

Let $\sigma_1 \leq \sigma_2 \leq \ldots \leq \sigma_n$ denote the n real eigenvalues of the symmetric matrix H, and let $m \leq n$ denote the multiplicity of the smallest eigenvalue σ_1, i.e., $\sigma_1 = \ldots = \sigma_m < \sigma_{m+1} \leq \ldots \leq \sigma_n$. Let $\Lambda = \text{diag}[\sigma_1, \ldots, \sigma_n]$. Let $q_1, \ldots, q_n$ denote an orthonormal set of eigenvectors of H corresponding to the eigenvalues $\sigma_1, \ldots, \sigma_n$ and let $Q = [q_1, \ldots, q_n]$, so $Q^T Q = I$ and $H = Q \Lambda Q^T$. Let $Q_m = [q_1, \ldots, q_m]$ be the matrix whose columns span the m-dimensional invariant subspace corresponding to the smallest eigenvalue σ_1 and let $Q_{n-m} = [q_{m+1}, \ldots, q_n]$.

The eigenvalues of $(H + \lambda^* I)$ in (6) are $(\sigma_1 + \lambda^*), \ldots, (\sigma_n + \lambda^*)$ and

$$(H + \lambda^* I) = Q(\Lambda + \lambda^* I)Q^T. \tag{13}$$

If $\sigma_1 > 0$, then H is positive definite, Δ is finite for the foregoing choice (12), the ball constraint is redundant and $z^* - x = -H^{-1}g$ is a direction of descent[2].

---

[2] This is the 'safe' case from a *mathematical* standpoint. Numerical issues are considered in Chapter 5.

Thus, henceforth in this subsection, we only consider the case

$$\sigma_1 \leq 0. \tag{14}$$

Since $(H + \lambda^* I) \geq 0$, we have $\sigma_1 + \lambda^* \geq 0$. Combined with (14), this gives

$$\lambda^* \geq -\sigma_1 \geq 0. \tag{15}$$

There are four cases to consider:

a) The simplest case arises when H is nonsingular and $Q_m^T g \neq 0$, i.e, the (negative) gradient vector has a component in the invariant subspace corresponding to the most negative eigenvalue. Δ is finite in (4) and (12). To find the optimal λ^*, let us consider the solution of the following *one-dimensional nonlinear equation*, which is motivated by the first two conditions of (5):

$$||w(\lambda)||_2 = \Delta, \quad w(\lambda) = -(H + \lambda I)^{-1} g. \tag{16}$$

The function $w(\lambda)$, in turn, can be explicitly defined in terms of the eigenvalues of H as follows:

$$||w(\lambda)||_2^2 = ||[Q(\Lambda + \lambda I)Q^T]^{-1} g||_2^2 = ||Q(\Lambda + \lambda I)^{-1} Q^T g||_2^2.$$

Thus

$$||w(\lambda)||_2^2 = \sum_{k=1}^{n} \frac{\gamma_k^2}{(\sigma_k + \lambda)^2} = \frac{\sum_{k=1}^{m} \gamma_k^2}{(\sigma_1 + \lambda)^2} + \sum_{k=m+1}^{n} \frac{\gamma_k^2}{(\sigma_k + \lambda)^2} \tag{17}$$

where $\gamma_k = (Q^T g)_k$, $k = 1, \ldots, n$.

Observe that the function $||w(\lambda)||_2$ is *strictly decreasing* over $(-\sigma_1, \infty)$ and that it is the sum of a finite collection of convex functions and is therefore also *convex*.

By assumption, $Q_m^T g \neq 0$, and the quantity $\sum_{k=1}^{m} \gamma_k^2$ in (17) is thus nonzero. Hence

$$\lim_{\lambda \to -\sigma_1^+} ||w(\lambda)||_2^2 = +\infty \quad \text{and} \quad \lim_{\lambda \to +\infty} ||w(\lambda)||_2^2 = 0.$$

Since Δ is finite, the equation

$$||w(\lambda)||_2 - \Delta = 0 \tag{18}$$

has a solution $\lambda^* \in (-\sigma_1, \infty)$, i.e., $\lambda^* > -\sigma_1 \geq 0$, the matrix $(H + \lambda^* I)$ is positive definite and $w(\lambda^*) = (z^* - x) = -(H + \lambda^* I)^{-1} g$ is a *direction of descent*.

The solution λ^* could be found inefficiently, but effectively, by a bisection algorithm. For more efficient techniques, see Chapter 4, Section 1.2.3.

b) Next, consider the case H singular and $Q_m^T g \neq 0$. The first assumption implies that $\Delta = \infty$ in (12) and thus the equation (18) has no solution in

45

$(-\sigma_1, \infty)$. We must have $\lambda^* = -\sigma_1$. Thus, $(H + \lambda^* I)$ is singular, i.e., positive semidefinite.

Whenever $\lambda > -\sigma_1$,

$$
\begin{aligned}
w(\lambda) &= -(H + \lambda I)^{-1} g \\
&= -Q(\Lambda + \lambda I)^{-1} Q^T g
\end{aligned}
$$

$$
= -[Q_m | Q_{n-m}]
\begin{bmatrix}
\sigma_1 + \lambda & & & & & \\
& \ddots & & & & \\
& & \sigma_1 + \lambda & & & \\
& & & \sigma_{m+1} + \lambda & & \\
& & & & \ddots & \\
& & & & & \sigma_n + \lambda
\end{bmatrix}^{-1}
\begin{bmatrix}
Q_m^T g \\
Q_{n-m}^T g
\end{bmatrix}
$$

$$
= -Q_m \frac{Q_m^T g}{(\sigma_1 + \lambda)} - Q_{n-m}
\begin{bmatrix}
\sigma_{m+1} + \lambda & & \\
& \ddots & \\
& & \sigma_n + \lambda
\end{bmatrix}^{-1}
Q_{n-m}^T g. \tag{19}
$$

Since $Q_m^T g \neq 0$, we see that as $\lambda \to -\sigma_1^+$, the direction $w(\lambda)$ approaches a direction of *negative curvature* corresponding to the most negative eigenvalue of H. Also, (19) implies that this limiting direction, say $w(\lambda^*)$, is a *direction of descent*, i.e., $w(\lambda^*)^T g < 0$.

c) Thirdly, consider the case H nonsingular and $Q_m^T g = \sum_{k=1}^{m} \gamma_k^2 = 0$. Now Δ is finite. Consider again the function $\|w(\lambda)\|_2$ defined by (17) when $\lambda > -\sigma_1$, which takes the form

$$
\|w(\lambda)\|_2^2 = \sum_{k=m+1}^{n} \frac{\gamma_k^2}{(\sigma_k + \lambda)^2} \equiv \bar{\Delta}(\lambda)^2. \tag{20}
$$

As $\lambda \to -\sigma_1^+$, the quantity $\bar{\Delta}(\lambda) \to \bar{\Delta}$, where $\bar{\Delta}$ is a finite number defined by

$$
\bar{\Delta}^2 \equiv \sum_{k=m+1}^{n} \frac{\gamma_k^2}{(\sigma_k - \sigma_1)^2}. \tag{21}
$$

There are two subcases to consider:

(i) If $\bar{\Delta} \geq \Delta$ then the nonlinear equation $\|w(\lambda)\|_2 - \Delta = 0$ has a solution $\lambda^* \in (-\sigma_1, \infty)$ and again the corresponding direction $w(\lambda^*)$ is a direction of descent.

(ii) However, Δ can be arbitrarily large, because H can be arbitrarily close to singularity. When $\bar{\Delta} < \Delta$ then the optimal value of λ^* is $\lambda^* = -\sigma_1 > 0$ (since H is nonsingular for the case under consideration). Now $(H + \lambda^* I)$ is only positive semidefinite.

Present assumptions imply that the first term in (19) is an indefinite quantity, i.e., $\frac{0}{0}$. One could obtain a (non-optimal) direction, say $p(\lambda^*)$, from (19)

46

by setting this the first term to zero, giving

$$p(\lambda^*) = -Q_{n-m} \begin{bmatrix} \sigma_{m+1} - \sigma_1 & & \\ & \ddots & \\ & & \sigma_n - \sigma_1 \end{bmatrix} Q_{n-m}^T g \qquad (22)$$

and again this is a direction of descent.[3] However, by assumption for the case under consideration, the second condition of (5) implies that the optimal solution $w(\lambda^*) = z^* - x$ of the model must satisfy $\|w(\lambda^*)\|_2 = \Delta$ (H is nonsingular so $\lambda^* = -\sigma_1 \neq 0$). This equation will generally not be satisfied by $p(\lambda^*)$ defined by (22).

To find a solution that is closer to the optimal solution, consider the singular system

$$(H + \lambda^* I)w(\lambda^*) = -g, \qquad (23)$$

corresponding to $\lambda^* = -\sigma_1$. Using (13) and $Q_m^T g = 0$, we can expand this equation as follows:

$$\begin{bmatrix} 0 & & & & & \\ & \ddots & & & & \\ & & 0 & & & \\ & & & \sigma_{m+1} - \sigma_1 & & \\ & & & & \ddots & \\ & & & & & \sigma_n - \sigma_1 \end{bmatrix} \begin{bmatrix} Q_m^T w(\lambda^*) \\ Q_{n-m}^T w(\lambda^*) \end{bmatrix} = - \begin{bmatrix} 0 \\ Q_{n-m}^T g \end{bmatrix}$$

Thus, the first m components of $Q^T w(\lambda^*)$, i.e. $Q_m^T w(\lambda^*)$, can assume any value, say w_m, and the remaining components are given by

$$w_{n-m} = - \begin{bmatrix} \sigma_{m+1} - \sigma_1 & & \\ & \ddots & \\ & & \sigma_n - \sigma_1 \end{bmatrix}^{-1} Q_{n-m}^T g.$$

Therefore

$$\begin{aligned} w(\lambda^*) &= Q \begin{bmatrix} w_m \\ w_{n-m} \end{bmatrix} \\ &= [Q_m | Q_{n-m}] \begin{bmatrix} w_m \\ w_{n-m} \end{bmatrix} \\ &= Q_m w_m - Q_{n-m} \begin{bmatrix} \sigma_{m+1} - \sigma_1 & & \\ & \ddots & \\ & & \sigma_n - \sigma_1 \end{bmatrix}^{-1} Q_{n-m}^T g. \end{aligned}$$

By assumption $g^T Q_m = 0$, which implies that $g^T w(\lambda^*) < 0$ for any choice of w_m. Thus w_m can be chosen so that $\|w(\lambda^*)\|_2 = \Delta$. (For example, take

---

[3] We assume $g \neq 0$. Then $Q_m^T g = 0$ and nonsingularity of the diagonal matrix in (22) imply that $Q_{n-m}^T g \neq 0$.

$w_m = \mu[1, 1, \ldots, 1]^T$ and choose $\mu = \sqrt{(\Delta^2 - \bar{\Delta}^2)/m}$. For the case under consideration, $\bar{\Delta} < \Delta$, so $\mu > 0$.) This also has the effect of including components of the eigenvectors corresponding to the most negative eigenvalue of H into the resulting direction of descent.

d) Finally, consider the case H singular and $Q_m^T g = 0$. Now $\Delta = \infty$ and we are in the foregoing case c) (ii). However, should we seek to satisfy the equation $\|w(\lambda^*)\|_2 = \Delta$, the optimal solution would be dominated by the columns of Q_m and we would not have a direction of descent. We must be content with the descent direction $p(\lambda^*)$ defined by (22). An alternative is to choose a large but finite value for Δ and proceed as in case c) (ii).

Thus, in all four cases we have found a suitable search direction of descent. *The heart of the procedure is an eigendecomposition of H and a zero-finding procedure to solve a nonlinear equation in one variable*, namely,

$$\|w(\lambda)\|_2 - \Delta = 0,$$

where $w(\lambda)$ is defined by (17).

2.4   Choosing the Next Iterate

A simple conceptual procedure based on the Newton model could use the foregoing direction of descent in a *line search* procedure in order to obtain a new iterate with a reduced function value, i.e.,

$$x_+ = \text{argmin}_{\alpha \geq 0} f(x + \alpha w(\lambda^*)). \tag{24}$$

Thus, in summary, a conceptual method based on the Newton model involves:

- the eigendecomposition of the Hessian matrix H,

- a strategy for choosing Δ at each iteration,

- a unidimensional zero-finding procedure to obtain the optimal Lagrange multiplier of the ball constraint together with a direction of descent, and

- a unidimensional minimizing routine to find an improving iterate along this direction.

It should be evident that the foregoing four aspects of the underlying method are *interrelated*, that there are *different ways of organizing a particular Newton algorithm* based on them, and that many *efficiency and robustness enhancing variants* are possible. Therefore, in contrast to the metric-based development of Chapter 2, we do not state a particular conceptual Newton algorithm here. The model-based Newton method and the variety of algorithms that can be derived from it, including so-called trust-region algorithms, will be discussed in more detail in Chapter 4, Section 1.2.

3 The Quasi-Newton Model

When the Hessian matrix H is unavailable or too expensive to provide, it must be approximated by an $n \times n$ symmetric matrix, say M. The matrix M can then be used in place of H in the foregoing model-based approach, i.e., a *Quasi-Newton Model* can be formulated as follows:

$$\text{minimize } g^T(z - x) + \frac{1}{2}(z - x)^T M(z - x) \tag{25}$$

s.t.

$$\|z - x\|_2 \leq \Delta. \tag{26}$$

For example, the quantity Δ could be chosen to be $\|M^{-1}g\|_2$, analogously to (12). A development that exactly parallels that of the Newton model of Section 2 can be carried out to obtain a direction of descent. It is important to note that a direction of negative curvature with respect to the Hessian M of the model (25) is *not necessarily a direction of negative curvature* for the function f. However, the search direction developed analogously to Section 2 would be *a direction of descent* for f at the current iterate in all cases.

Suppose that the next iterate, obtained using this direction of descent in a line search procedure[4], is x_+ with corresponding gradient g_+, and, as before, let $s = x_+ - x$ and $y = g_+ - g$. The item of immediate interest to us here is the revision or updating of M using the information s and y gathered at the current iteration.

A standard mean value theorem for vector valued functions states that

$$\left[\int_0^1 H(x + \theta s)d\theta \right] s = y, \tag{27}$$

i.e., the averaged Hessian matrix over the current step transforms the step s into the corresponding gradient change y. In revising M in order to reflect new information gathered over the current iteration, it is natural to impose the requirement that the updated matrix, say M_+, have the same property, the so-called *quasi-Newton* or *Secant* relation:

$$M_+ s = y. \tag{28}$$

Note that M_+ is *not* required to be positive definite. The Variable-Metric relation of Chapter 2, Section 4.1 is the Quasi-Newton or Secant relation with the added condition $M_+ > 0$ imposed on (28).

3.1 The Symmetric Rank One (SR1) Update

The Symmetric Rank One (SR1) Update makes the simplest possible modification to M, adding to it a matrix γuu^T, where γ is a real number and u is a

---

[4]Instead, a trust-region approach could be used, as discussed in Chapter 4, Section 2.2.3.

vector in $\Re^n$. The unique matrix of the form $M_+ = M + \gamma uu^T$ that satisfies (28) is

$$M_+ = M + \frac{(y - Ms)(y - Ms)^T}{(y - Ms)^T s}. \tag{29}$$

We have encountered this update in Chapter 1, and its derivation is identical to that of (1:46) after we identify M_j with M and M_{j+1} with M_+. As noted in Chapter 1, Section 7.1, the derivation of the SR1 update does not depend on the quadratic context, and the update exists provided $(y - Ms)^T s \neq 0$. If this quantity is zero, or more safely, if $|(y - Ms)^T s| \leq C \|y - Ms\|_2 \|s\|_2$, for some small constant $C > 0$, then the update can be skipped, i.e., M_+ can be set to M.

Having updated M to M_+, a model analogous to (25)-(26) can be developed at x_+ and the procedure repeated.

In marked constrast to the convex quadratic case considered in Chapter 1, Section 7.1, we can no longer expect the hereditary quasi-Newton property to hold for arbitrary smooth functions, i.e., if s_- denotes a prior step with corresponding gradient change y_- then, in general, $M_+ s_- \neq y_-$. Additional conditions must be imposed on these quantities to obtain the hereditary property on arbitrary functions. For example, we can state the following result:

Let s and s_- be linearly independent (hence non-zero) steps. Then there exists a symmetric matrix M_+ such that

$$M_+ s = y \text{ and } M_+ s_- = y_- \tag{30}$$

if and only if

$$s^T y_- = y^T s_-. \tag{31}$$

The proof is as follows:

To establish necessity, suppose (30) holds with M_+ a symmetric matrix. Then $s_-^T M_+ s = s_-^T y$ and $s^T M_+ s_- = s^T y_-$. Since M_+ is symmetric, condition (31) is satisfied.

To establish sufficiency, suppose (31) holds. We will construct a suitable symmetric matrix M_+ satisfying (30).

Let M_- be any symmetric matrix, for example, $M_- = I$. Let

$$M = M_- + \frac{(y_- - M_- s_-)u_-^T + u_-(y_- - M_- s_-)^T}{u_-^T s_-} - \frac{(y_- - M_- s_-)^T s_-}{(u_-^T s_-)^2} u_- u_-^T, \tag{32}$$

where u_- is an arbitrary vector such that $u_-^T s_- \neq 0$. It can be verified directly that $Ms_- = y_-$. Furthermore, the choice $u_- = (y_- - M_- s_-)$ would make (32) equivalent to the symmetric rank-one update, and we can make this simpler update in the construction whenever $(y_- - M_- s_-)^T s_- \neq 0$.

Now we repeat the procedure. Thus, pick any vector u such that $u^T s \neq 0$ and additionally such that $u^T s_- = 0$. Since s and s_- are linearly independent by assumption, such a vector u always exists. Let

$$M_+ = M + \frac{(y - Ms)u^T + u(y - Ms)^T}{u^T s} - \frac{(y - Ms)^T s}{(u^T s)^2} uu^T. \tag{33}$$

Again, $M_+ s = y$. From (33), $u^T s_- = 0$ and $Ms_- = y_-$,

$$M_+ s_- = Ms_- + ((y - Ms)^T s_-)u = y_- + ((y - Ms)^T s_-)u.$$

But $(y - Ms)^T s_- = y^T s_- - s^T Ms_- = y^T s_- - s^T y_- = 0$, using the sufficiency assumption. Thus $M_+ s_- = y_-$, and hence (30) holds. $\square$

The foregoing result can be generalized in a straightforward way: there exists a symmetric matrix, say M_{k+1}, such that $M_{k+1} s_i = y_i$ for a set of step and gradient-change pairs $(s_i, y_i), i = 1, \ldots, k$ if and only if $s_i^T y_j = s_j^T y_i \ \forall \ i \neq j$. Also, (32) or (33) are examples of symmetric rank-two updates that have greater flexibility than the SR1-Update[5].

The SR1-Update and algorithms that can be derived from it are further discussed in Chapter 4, Section 2.2.

4 The CG-Related Model

The CG-metric of Chapter 3, Section 5 was obtained by reinitiating the B-Update at each iteration. This suggests an analogous model-based development that reinitiates the SR1-Update.

Thus, after the matrix M has been used to obtain a new iterate x_+ and associated step and gradient-change pair (s, y), suppose that it is replaced by the identity matrix I. Then the update (29) over (s, y) becomes

$$M_+ = I + \frac{(y - s)(y - s)^T}{(y - s)^T s}, \tag{34}$$

where we assume $(y - s)^T s \neq 0$. The matrix M_+ can be represented *implicitly* by storing the vectors s and y that define it, *so no storage of matrices is needed*.

The associated CG-related model at the new iterate x_+ is given by

$$\text{minimize } g_+^T(z - x_+) + \frac{1}{2}(z - x_+)^T M_+(z - x_+) \tag{35}$$

s.t.

$$\|z - x_+\|_2 \leq \Delta_+ = \|M_+^{-1} g_+\|_2. \tag{36}$$

Considerable simplifications occur as a result of the special form of M_+.

Thus, finding the eigendecomposition of M_+ is trivial. Is eigenvalues are $1 + \frac{(y-s)^T(y-s)}{(y-s)^T s}$ and 1, the latter with multiplicity $(n - 1)$. In particular, M_+

---

[5] See also Chapter 6, Section 2.1.1.

is singular when $(y - s)^T y = 0$. The corresponding normalized eigenvectors are $(y - s)/\|y - s\|_2$ and $n - 1$ orthonormal vectors spanning the orthogonal complement of $y - s$.

Procedures for finding the optimal Lagrange multiplier of the ball constraint in the foregoing model and the associated direction of descent, discussed in Section 2, will also specialize considerably. For example, the optimal Lagrange multiplier can be shown to be the solution of a low-degree polynomial and can be found explicitly.

These topics and the viability of the approach remain to be explored.

5   The Cauchy Model

Finally, we come to the trivial case, when the Hessian of the model is always taken to be the identity matrix I. The corresponding model at x is

$$\text{minimize } g^T (z - x) + \frac{1}{2}(z - x)^T (z - x) \tag{37}$$

s.t.

$$\|z - x\|_2 \leq \Delta. \tag{38}$$

The optimal direction $z^* - x$ is along the direction of steepest descent $-g$ for *any* choice of Δ, i.e., the ball constraint is always redundant. This is the complementary model-based description of the steepest-descent method of Chapter 2, Section 2.

More generally, one can define a *Cauchy* model as follows:

$$\text{minimize } g^T (z - x) + \frac{1}{2}(z - x)^T D^2 (z - x) \tag{39}$$

s.t.

$$\|z - x\|_2 \leq \Delta = \|D^{-2} g\|_2 \tag{40}$$

where D is a *diagonal* matrix with *positive* diagonal elements. Again the ball constraint is redundant for the given choice of Δ and we have the complementary (and mathematically equivalent) model-based description of the Cauchy metric-based method of Chapter 2, Section 3.

One can envision an extension of the foregoing model, which uses a diagonal matrix for the Hessian of the model with elements obtained by finite difference approximations. For example, when gradients are estimated by central differences then the same function values could be used to estimate diagonal Hessian elements. These elements could be of either sign and the local approximating model would have to be constrained when any elements are negative. We could term such an approach a *Modified-Cauchy Method* and it too has not been explored to date. Its computational significance is unclear, but nevertheless it should be explicitly identified in any overall schemata.

6 Notes

Section 2: The development draws substantially on Moré and Sorensen [1982] and Moré [1983], but differs in that the method described in this chapter relies directly on a line search.

CHAPTER 4
THE NEWTON-CAUCHY FRAMEWORK

A simple and elegant picture emerges from the complementary perspectives of Chapters 2 and 3 to which the motivation and background of Chapter 1 lend added dimension. We bring this picture more clearly into focus in this chapter.

Our development will now be much more concise, in contrast to the first three chapters of the monograph, which were detailed and relatively self-contained. Proofs of assertions will only be outlined, and we will make extensive reference to other published research articles. Our aim is to provide the reader with *an overall framework for the main categories of conceptual methods*[1] as summarized in Figure 4.1, and a *staging-ground* for the development of new methods.

$Metric - Based$ $Methods$	$Model - Based$ $Methods$
$Modified - Newton$	$Newton$
$Variable - Metric$ $Quasi - Newton$	$Secant$ $Quasi - Newton$
$CG - related$ $Mod. - Newton$ and VM	$CG - related$ $Newton$ and $Secant$
$Cauchy$	$Modified - Cauchy$

Figure 4.1

With reference to Figure 4.1, suppose that the Hessian matrix H or Hessian approximation M is *positive definite* in a model-based approach. Let $\bar{H}$ denote either of these two quantities. As usual, let g denote the gradient vector at the current iterate. A reconditioner, say $\bar{R}$, for a conceptually equivalent metric-based description can be obtained from *any* factorization of the form $\bar{H} = \bar{R}^T \bar{R}$. Conversely, when provided with a nonsingular reconditioner $\bar{R}$ for a metric-based approach, a conceptually equivalent model-based formulation can be obtained, by using the positive-definite matrix $\bar{R}^T \bar{R}$ for the Hessian of the quadratic model's objective function. Recall that the ball constraint with

---

[1] A conceptual method is the set of basic ideas or principles that collectively define a particular computational approach.

$\Delta = \|\bar{H}^{-1}g\|_2$ given by (3:12) is redundant in the basic model-based formulation when the model's Hessian is positive-definite, i.e., the model is then effectively unconstrained. Thus, whenever $\bar{H} = \bar{R}^T \bar{R} > 0$, the model-based Newton (unconstrained, positive-definite Hessian) approach and metric-based Cauchy approach are *mathematically equivalent*; we shall adopt the view that that they are two different ways of describing the *same* conceptual method, which we shall henceforth identify by the label: *metric-based*.

It is important to note that metric-based and unconstrained model-based formulations may be conceptually equivalent, yet quite different when viewed from the standpoint of how calculations are organized i.e., they may have different numerical properties and offer rather different opportunities for modification to enhance numerical stability. For example, compare Procedure VM/B of Chapter 2, Section 4.2 and a variant of Procedure QN/B of Chapter 1, Section 8, which develops approximations W_j to the inverse Hessian (instead of M_j) and substitutes an exact line search for the formula for the step-length α_j^*, so the modified QN/B procedure can be applied to an arbitrary smooth function. From a mathematical standpoint, these two procedures are identical, but they differ substantially from a numerical standpoint, because the computations are organized in very different ways. In particular, procedure VM/B is much more amenable to numerical stabilization when the reconditioner is maintained in upper triangular form. This point will be taken up again in the next chapter, which addresses numerical implementation issues. See also the discussion in Chapter 1, Section 11 and Chapter 6, Section 2.1.2.

The second important point to note with reference to Figure 4.1 is that the model-based approach subsumes the metric-based approach, because the former is also valid when the Hessian or approximation to the Hessian matrix is indefinite. In this case there is no associated factorization of $\bar{H}$ that can be used to define a reconditioner $\bar{R}$ for a complementary metric-based formulation.

1 Newton Methods

An interesting historical account of the development of the original *unidimensional* Newton and secant methods is given by Ypma [1992]. Nowadays, in the optimization context, Newton's method is used as a *generic* name for multidimensional minimization techniques that employ second-order derivative information, including the case when this information is estimated by differences of lower-order derivatives or function values, over numerically infinitesmal steps.

1.1 Metric-Based Newton

Let H denote the Hessian matrix $\nabla^2 f(x)$ at the current iterate x. When H is positive definite, a descent search direction d is defined by solving the system $Hd = -g$, where g is the gradient vector of f at x. A line search procedure is initiated at x along the descent direction d, in order to obtain an improving point.

When H is indefinite, it is altered or modified in order to obtain a positive-definite approximation $\bar{H}$ in one of several possible ways:

- Form the eigendecomposition of H and reverse the signs of the negative eigenvalues or modify them in some other suitable manner, see, for example, Greenstadt [1967].

- Initiate the Cholesky factorization of H. If this procedure fails at some intermediate step because H is not positive definite, then modify the offending diagonal element of H so as to ensure that the Cholesky procedure can be continued. The resulting factors are those of a positive definite modification, say $\bar{H}$, of the original matrix H, see Gill and Murray [1974].

- Form the Bunch and Parlett [1971] factorization of $H = LDL^T$ where L is a unit lower triangular matrix and D is block diagonal with 1×1 or 2×2 matrices on its diagonal. Then modify any 2×2 elements of D so that the resulting matrix, say $\bar{D}$, is positive definite. Let $\bar{H} = L\bar{D}L^T$.

- Add a matrix λD to H, where D is a diagonal matrix with positive diagonal elements (typically $D = I$) and $\lambda > 0$ is is taken large enough to ensure that $(H + \lambda D) > 0$, see Levenberg [1944], Marquardt [1963].

The foregoing methods are all *ad-hoc* modifications of H designed to obtain a positive-definite matrix $\bar{H}$ and its factors. A descent search direction is then again obtained by solving $\bar{H}d = -g$, using the matrix factors of $\bar{H}$. The resulting method is called a *Modified-Newton Method*.

As noted in the introduction to this chapter, a conceptually equivalently formulation can be given, a posteriori, by defining a reconditioned steepest-descent direction, where the reconditioner is obtained from the factorization of $\bar{H}$. Thus, *a modified-Newton method is essentially metric-based*.

At a strict local minimizing point x^*, the Hessian matrix $H^* = \nabla^2 f(x^*)$ is positive definite, and no modification of the Hessian will be needed when the foregoing procedure is initiated from a starting point *sufficiently close* to x^*. Asymptotically, Newton's method will converge at a quadratic rate (see, for example, Luenberger [1984]). Convergence of a modified-Newton method from an *arbitrary* starting point is discussed in Chapter 5.

1.2 Model-Based Newton

When applied to a non-convex function f, whose Hessian matrix H is not everywhere positive definite, the foregoing modified-Newton methods are *ad hoc* in character. The more natural setting for Newton's method is the model-based approach with a step-size (trust-region) constraint as introduced in Chapter 3. As described there, the method in its basic form involves:

- an orthogonal *factorization* of $H = \nabla^2 f(x)$.

- a strategy for defining the *ball-constraint*.

- a zero-finding routine to obtain the optimal Lagrange *multiplier* of the ball constraint.

- the extraction of a direction of descent followed by a line search to find an *improving point*.

There are different ways of organizing an algorithm around these basic ideas, and variants on the foregoing approach address each of its basic components (italicized), in order to enhance efficiency and robustness. Let us consider each one in turn.

It is important to note that the four aspects of the method discussed below are *distinct, but interrelated*. Decisions taken with regard to one will influence decisions that are or can be taken with regard to the others.

1.2.1 Factorization

One should not necessarily shrink from forming the eigendecomposition of H, especially on problems where n is of small or medium dimension. The eigendecomposition can nowadays be computed very efficiently by modern techniques of computational linear algebra (see Parlett [1980]). However, this operation must be done each time a new iterate x and its associated Hessian matrix H are generated, and thus the opportunity offers itself of being able to derive cumulative gains by using a cheaper matrix factorization at each iteration.

The definitive articles in this area are Moré [1983] and Moré and Sorensen [1982]. The variants discussed there revert to the form (3:16), and obtain values of the unidimensional function $w(\lambda)$ for any given $\lambda \in (-\sigma_1, \infty)$, by solving the system of equations $(H + \lambda I)w(\lambda) = -g$ *using the Cholesky factorization* of $H + \lambda I$.

When operations involved in the Newton algorithm require the solution of a system of equations defined by an indefinite matrix, for example, the computation of the quantity Δ by (3:12), then the Bunch and Parlett [1971] factorization, a natural generalization of the Cholesky factorization, can be profitably employed. See also Sorensen [1977].

1.2.2 Ball-Constraint

It is possible to use a more general ellipsoidal constraint defined by $(z-x)^T D(z-x) \leq \Delta$ where D is a diagonal (scaling) matrix with positive diagonal elements. Extensions of the basic procedure are straightforward; see, for example, Moré [1983]. Strategies for choosing the ball-constraint radius Δ will be considered in Section 1.2.4.

1.2.3 Multiplier

A good approximation $\bar{\lambda}$ to λ^* with $\bar{\lambda} > \max\{0, -\sigma_1\}$ will suffice, i.e., the optimal Lagrange multiplier itself does not have to be computed. Thus opportunities abound for greatly enhancing efficiency.

Effective techniques have been suggested by Hebden [1973], Reinsch [1971], and Gay [1981], and a definitive synthesis of these approaches is given by Moré [1983] and Moré and Sorensen [1982]. The main idea is as follows:

The function $\|w(\lambda)\|_2 - \Delta$ in (3: 16) and (3:17) has a pole at $\lambda = -\sigma_1$, which slows down the convergence of unidimensional zero-finding algorithms. Therefore, formulate *the inverse system*:

$$\frac{1}{\|w(\lambda)\|_2} - \frac{1}{\Delta} = 0, \tag{1}$$

and solve it using the unidimensional Newton's method (assume $g \neq 0$ when evaluating $w(\lambda)$). The function on the left-hand side of (1) has no poles and tends to be nearly linear on $(-\sigma_1, \infty)$, see Reinsch [1971]. Thus global convergence of the unidimensional Newton's method is assured, and a good approximation to the solution is usually obtained in two or three iterations, each requiring an evaluation of $w(\lambda)$, i.e., a factorization of $H + \lambda I$. For full details, consult the foregoing references.

1.2.4 Improving Point

There are two basic approaches:

- *Line Search Approach:* The underlying philosophy is to define the radius Δ of the ball constraint in such a way as to be redundant when H is positive definite. When this occurs, the method becomes the standard metric-based Newton's method. A simple strategy is to choose Δ by (3:12), and this quantity can be safeguarded, if desired, by placing some upper bound, say Δ_{max}, on the step-size. When Δ is always restricted to be finite, only a subset of the cases considered in Chapter 3, Section 2.3 can arise, and the resultant procedure for obtaining a direction of descent simplifies. However, an advantage of permitting Δ to take on the value infinity is that a direction of negative curvature, corresponding to the most negative eigenvalue, is then selected in a natural way.

  Note also that once the optimal Lagrange multiplier or an approximation to it, say $\bar{\lambda}$, is available, a metric-based procedure can be defined *a posteriori*, whenever $(H + \bar{\lambda} I)$ is positive definite, by factorizing this matrix and using its factor to define a reconditioner.

  The line search approach has not been fully explored to date. For some further discussion of convergent implementable algorithms derived from it, see the next chapter.

- *Trust Region Approach*: Here the philosophy is to 'trust' the model only in a localized region around the current iterate, and to treat Δ as an explicit, always finite, dynamically-controlled parameter that 'drives' the algorithm. (Strategies for changing Δ are often ad-hoc, but one step of the strategy for updating the parameter resembles one step of a line-search procedure.) The method is very robust, but has the drawback that it is not mathematically equivalent to the standard Newton method when H is positive definite. For example, if f is a strictly convex quadratic function and the initial trust region radius Δ is too small, the algorithm does not converge in a single step.

Definitive studies are given, for example, by Moré and Sorensen [1982], Sorensen [1982a,b], Moré [1983], Burke, Moré and Toraldo [1990].

2   Quasi-Newton Methods

The seminal ideas on which Quasi-Newton methods rely are due to Davidon [1959] (see also Anderson, Davidon, Glicksman and Kruse [1955] and Anderson and Davidon [1957]). They were brought to the attention of the optimization community in the important clarification and promulgation of Fletcher and Powell [1963], and numerous researchers have since contributed to the full-fledged development and implementation of the method.

The discussion of Chapters 2 and 3 shows that the metric-based B-update and the model-based SR1 update are both natural choices, each in its own context. Recall that the SR1 update makes the simplest natural modification of the symmetric matrix M, namely, a symmetric rank-1 augmentation $\gamma uu^T, \gamma \in \Re, u \in \Re^n$, subject to satisfying the secant (QN) relation. This yields the unique update (3:29). Likewise, the B-update makes the simplest natural modification of the nonsymmetric, nonsingular reconditioner R, namely, a general rank-1 augmentation $uv^T, u \in \Re^n, v \in \Re^n$, subject to satisfying the variable metric (QN) relation. Because there is still room for choice, it seeks an updated reconditioner R_+ that is as close as possible to R, yielding the update (2:49). Observe that the augmentation $R + uv^T$ could equally well be written in the multiplicative form $(I + uv^T)R$, without loss of generality. If instead one had employed the more restrictive modification $(I + \gamma uu^T)J$, where $\gamma \in \Re, u \in \Re^n$, and required the updated reconditioner to satisfy the variable metric relation, then it is easily verified that the updated reconditioner R_+ corresponding to the SR1 update would be obtained, *whenever it exists*, i.e., whenever the SR1 update of the positive definite matrix $M = R^T R$ yields a positive definite matrix $M_+ = R_+^T R_+$. A symmetric modification of the generally nonsymmetric matrix R is, of course, not a natural choice, leading to the conclusion that the SR1 update is not a natural candidate *in a metric-based setting*.

In Chapter 1, we introduced the B-update and the SR1 update in the specialized setting of minimizing a quadratic function. In order to highlight the respective strengths and weaknesses of the two updates, consider a smooth function that is nonconvex over part of its domain containing the starting point and (albeit artificially) *a strictly convex quadratic,* as in Chapter 1, in some level set containing the minimizing point. Suppose the true indefinite Hessian matrix, say H_0, is available to a quasi-Newton method at the starting point x_0. The BFGS variable-metric method (based on the B-Update) must be initiated with a modification of H_0 that produces a positive definite matrix, say $\bar{H}_0$. When it enters the quadratic domain and line searches are exact, it will converge in at most n further steps. The SR1 secant method, on the other hand, can be initiated with the true Hessian H_0, iterates would eventually enter the quadratic region, and thereafter the SR1 method would normally converge in at most n further steps. Note that exact line searches would not be needed.

Quadratics are useful for gaining insight into optimization methods, but, of course, too much weight should not be attached to behaviour in this specialized setting. Good behaviour on a quadratic does not necessarily imply good behaviour in general, but poor performance on a quadratic is often indicative of poor performance on more general functions.

We will now discuss the two basic classes of QN methods in more detail.

2.1 Metric-Based QN

The B-Update of a matrix $M > 0$ over a step s, with associated gradient change y and $y^T s > 0$, is given by (2:51), namely,

$$M_+ = M - \frac{(Ms)(Ms)^T}{s^T Ms} + \frac{yy^T}{y^T s}, \tag{2}$$

and the update of the inverse $W = M^{-1}$ is

$$W_+ = \left(I - \frac{sy^T}{y^T s}\right) W \left(I - \frac{sy^T}{y^T s}\right)^T + \frac{ss^T}{y^T s}. \tag{3}$$

The B-Update was arrived at independently, and in different ways, by Broyden [1970], Fletcher [1970], Goldfarb [1970] (with antecedents in Greenstadt [1970]), and Shanno [1970], and is thus popularly called the BFGS-Update. It is founded on the basic principles originated by Davidon [1959] and has proved to be one of the most effective in practice.

We now discuss the properties of this update and the resulting method.

2.1.1 Invariance of B-Update

The B-Update is *invariant under transformation of variables,* which can be seen as follows:

Let $x_j, g_j, R_j, M_j, s_j, y_j,\ j = 1, 2, \dots$ denote the generated sequence of iterates, gradients, reconditioners, Hessian approximations, steps and gradient changes. Let the initial Hessian approximation be $M_1 = P^T P$. Let us make the *fixed* transformation of variables or preconditioning $\tilde{x}[P] = Px$ and henceforth work in the $\tilde{x}[P]$-space. (It is convenient to explicitly identify the preconditioner in the transformed space.) The gradient g and Hessian H at any point x transform to $\tilde{g}[P] = P^{-T} g$ and $\tilde{H}[P] = P^{-T} H P^{-1}$, which can be verified by the chain rule, just as in Chapter 2, Section 4. Steps transform like points, i.e., $\tilde{s}_j[P] = Ps_j$ and gradient changes like gradients, i.e., $\tilde{y}_j[P] = P^{-T} y_j$. Hessian approximations will transform like the Hessian H, so $\tilde{M}_j[P] = P^{-T} M_j P^{-1}$, and, in particular, $\tilde{M}_1[P] = P^{-T}(P^T P)P^{-1} = I$. Reconditioners transform[2] like the Hessian factors, i.e., $\tilde{R}_j[P] = R_j P^{-1}$, and, in particular, $\tilde{R}_1[P] = PP^{-1} = I$.

Consider the B-Update in the *transformed* space:

$$\tilde{M}_{j+1}[P] = \tilde{M}_j[P] - \frac{(\tilde{M}_j[P]\tilde{s}_j[P])(\tilde{M}_j[P]\tilde{s}_j[P])^T}{\tilde{s}_j[P]^T \tilde{M}_j[P]\tilde{s}_j[P]} + \frac{\tilde{y}_j[P]\tilde{y}_j[P]^T}{\tilde{y}_j[P]^T \tilde{s}_j[P]}.$$

and $\tilde{M}_1[P] = I$.

Substituting for these quantities, we obtain

$$P^{-T} M_{j+1} P^{-1} = P^{-T}\left[M_j - \frac{(M_j P^{-1} Ps_j)(M_j P^{-1} Ps_j)^T}{s_j^T P^T P^{-T} M_j P^{-1} Ps_j} + \frac{y_j y_j^T}{y_j^T P^{-1} Ps_j} \right] P^{-1}.$$

Hence,

$$M_{j+1} = M_j - \frac{(M_j s_j)(M_j s_j)^T}{s_j^T M_j s_j} + \frac{y_j y_j^T}{y_j^T s_j}.$$

and $M_1 = P^T P$.

Observe that the updating relation is *invariant* under the transformation of variables defined by P. This is a key property. Note also that the development for quadratics in Chapter 1, Section 10 is simply a special case of the foregoing result.

We could have also arrived at the foregoing result by verifying, in a completely analogous way, that the updating relation of the *reconditioning factor* R_j in (2:49) is invariant. Then the updating relation for $M_j = R_j^T R_j$ must also be invariant.

2.1.2 Complementary B-Update

In the subsequent discussion, we will need to consider just two successive updates, so we can omit subscripts.

As usual, denote a reconditioner by R, the associated approximation to the Hessian matrix by $M = R^T R$, and its inverse by $W = M^{-1}$. We assume

---

[2] in other words, reconditioners can be preconditioned, illustrating the distinction between these two terms.

throughout that R is nonsingular, i.e., M is positive definite and W exists. Denote the current step by s and its corresponding gradient change by y. Assume that some line search is used to guarantee that $y^T s > 0$. Let

$$a = s^T M s, \quad b = y^T s, \quad c = y^T W y.$$

The foregoing assumptions imply that all three quantities are positive. For reasons discussed in Chapter 1, Section 11, it was initially popular to form approximations W to the inverse Hessian, and to require that the updated matrix W_+ satisfies the QN relation in the form

$$W_+ y = s. \tag{4}$$

This may be obtained from $M_+ s = y$ by making the interchanges

$$M \leftrightarrow W, \ M_+ \leftrightarrow W_+, \ s \leftrightarrow y. \tag{5}$$

This operation is called *complementing* or *dualizing* of an update, although the latter term is undesirable, because of its conflict with the conventional use of the term 'dual' in mathematical programming. It should be clear that the formulation analogous to that of the B-Update in Chapter 1, Section 7.2 (which as noted there applies to arbitrary differentiable functions) can be carried out with these interchanges. This is also true of the more general formulation in Chapter 2, Section 4.2, (with R used to denote the factor of the inverse Hessian). This leads to the Complementary B-Update

$$W_+ = W - \frac{(Wy)(Wy)^T}{c} + \frac{ss^T}{b}. \tag{6}$$

This was the original update proposed by Davidon [1959] and clarified by Fletcher and Powell [1963], and thus popularly called the DFP-Update. It can conveniently be abbreviated to the D-Update. Its inverse is given by complementing (3), namely,

$$M_+ = \left(I - \frac{ys^T}{b}\right) M \left(I - \frac{ys^T}{b}\right)^T + \frac{yy^T}{b}.$$

Note that the complementing operation is purely a symbol manipulation, and that any updating formula satisfying $M_+ s = y$ can be complemented to obtain a formula satisfying $W_+ y = s$. Indeed, the B-update was originally derived from the D-Update in this way by Fletcher [1970], and the complementing operation often provides a short-cut for proving properties about an update. For example, the formulae complementary to (1:56) is valid for the D-Update on a convex quadratic function. Thus the invariance property of the B-Update given at the end of Chapter 1, Section 7.2 is also true for the D-Update.

2.1.3 B-Family

A family of updates of M to $M_+(\mu)$ that broadens the B-Update was defined by Broyden [1970][3] as follows:

$$M_+(\mu) = M_+(0) + (\mu a)mm^T, \qquad (7)$$

where $\mu \in \Re$ is a parameter, the matrix $M_+(0)$, which forms the root of the family, is the B-Update (2) with the parameter value $\mu = 0$ explicitly attached, i.e.,

$$M_+(0) = M - \frac{Mss^T M}{a} + \frac{yy^T}{b}, \qquad (8)$$

and

$$m = \frac{y}{b} - \frac{Ms}{a}. \qquad (9)$$

Note that $m^T s = 0$ and thus $M_+(\mu)s = y \; \forall \; \mu$. Just as with the B-Update, all updates in the one parameter family are easily seen to be invariant with respect to transformations of variables, in the sense of Section 2.1.1.

The Complementary B-Family of updates of W to $W_+(\nu)$, depending on a parameter $\nu \in \Re$, is defined by

$$W_+(\nu) = W_+(0) + (\nu c)ww^T, \qquad (10)$$

where $W_+(0)$, the root of the family, is the Complementary B-Update (or D-Update), namely,

$$W_+(0) = W - \frac{Wyy^T W}{c} + \frac{ss^T}{b} \qquad (11)$$

and

$$w = \frac{s}{b} - \frac{Wy}{c}. \qquad (12)$$

Note that the family can be obtained by applying the complementing operation of the previous subsection to (7). Note also that (5) implies that a is complementary to c, b is unchanged, and m is complementary to w. The quantity $\nu \in \Re$ defines the complementary parameterization.

When $\nu = 1$, verify directly that (10) becomes

$$W_+(1) = W_+(0) + cww^T = \left(I - \frac{sy^T}{b}\right) W \left(I - \frac{sy^T}{b}\right)^T + \frac{ss^T}{b}, \qquad (13)$$

and observe that this is (3), the inverse form of the B-Update. Sometimes, it is mathematically convenient to parameterize the family (10) with the B-Update (13) used as the root rather than the D-Update, i.e., in the equivalent form:

$$W_+(\nu) = W_+(1) + (\bar{\nu}c)ww^T, \qquad (14)$$

---

[3] Note, *not* Broaden [1970].

where $\bar{\nu} \equiv (\nu - 1) \in \Re$. We shall find a use for this particular parameterization in the next subsection.

Observe that $ac = b^2$ if and only if $y \parallel Ms$ or, equivalently, $m = w = 0$. The Broyden families (7) and (10) reduce to $M_+(0)$ and $W_+(0)$, respectively, which, in this case, must be equivalent to the SR1 update. Henceforth, we can assume that $b^2 \neq ac$, which implies that $b^2 < ac$ by a version of the Cauchy-Schwartz inequality.

The values of μ and ν for which $M_+(\mu)$ and $W_+(\nu)$, respectively, are singular are

$$\mu^* = \nu^* = b^2/(b^2 - ac). \tag{15}$$

This can be deduced by noting that the determinant of $M_+(\mu)$ vanishes when the matrix is singular. Express the update in the form $M^{-1}(I + S)$, where S is a sum of matrices of rank 1, and use the relation

$$\det(I + u_1 u_2 + u_3 u_4) = (1 + u_1^T u_2)(1 + u_3^T u_4) - (u_1^T u_4)(u_2^T u_3).$$

Note that $\mu^* < 0$ under the previously mentioned assumption $b^2 \neq ac$. For metric-based updates, we are interested in the case

$$\mu > \mu^*, \tag{16}$$

and correspondingly $\nu > \nu^*$.

Given $M = W^{-1}$, a relation between the parameters μ and ν which ensures that $M_+(\mu)$ and $W_+(\nu)$ are inverses of one another is as follows:

$$\mu + \nu - \mu\nu(1 - ac/b^2) = 1. \tag{17}$$

This can be verified by directly multiplying the two matrices $M_+(\mu)$ and $W_+(\nu)$, with μ and ν related by (17), followed by a little tedious algebra.

Note that (17) is satisfied by $\mu = 0$ and $\nu = 1$, and viceversa. Thus the *inverse* of $M_+(0)$ is $W_+(1)$, i.e. the choice $\nu = 1$ in (10) gives the B-Update of the inverse Hessian approximation W. On the other hand, the *complement* of $M_+(0)$, as noted earlier, is $W_+(0)$, namely, the D-Update.

2.1.4 B-Family with Exact Line Searches

Consider a version of Procedure VM/B of Chapter 2, Section 4.2, again stripped bare of unnecessary detail, which updates approximations M_j to the Hessian instead of the reconditioner R_j and is extended to use any update in the one-parameter family (7). Note that the choice $\mu = 0 \,\forall\, j$ will make this extended conceptual procedure *mathematically equivalent* to Procedure VM/B.

Procedure QN/μ:

Given a starting point x_1, gradient g_1 and initial Hessian approximation M_1:

for $j = 1, 2, \ldots$.

Solve $M_j d_j = -g_j$ for d_j.

$\alpha_j^* = \mathrm{argmin}_{\alpha \geq 0} f(x_j + \alpha d_j)$.

$x_{j+1} = x_j + \alpha_j^* d_j$. Evaluate g_{j+1}.

$s_j = x_{j+1} - x_j$; $y_j = g_{j+1} - g_j$; $a_j = s_j^T M_j s_j$; $c_j = y_j^T M_j^{-1} y_j$; $b_j = y_j^T s_j$;
$\mu_j^* = b_j^2 / (b_j^2 - a_j c_j)$.   (If $b_j^2 = a_j c_j$ take $\mu_j^* = -\infty$.)

$m_j = \frac{y_i}{b_j} - \frac{M_j s_j}{a_j}$.

$M_{j+1} = M_j - \frac{M_j s_j s_j^T M}{a_j} + \frac{y_j y_j^T}{b_j} + (\mu_j a_j) m_j m_j^T$, where $\mu_j > \mu_j^*$.

end

Let us assume that exact line searches in the foregoing procedure are *unambiguously* resolved. We then have the following results:

Dixon's Theorem: The sequence of iterates developed by Procedure QN/μ is *independent* of the choice μ_j made at each iteration.

Powell's Theorem: Suppose $\mu_j = 0$ at iteration j, where $j \geq 2$, i.e., suppose the B-Update is used at the j'th iteration. Then M_{j+1} is *independent* of the choice made for $\mu_1, \ldots, \mu_{j-1}$.

The proofs of these two closely interrelated results are not difficult. Briefly, they can be outlined as follows: For mathematical convenience, work with the equivalent parameterization (14), which has the B-Update for its root. At iteration $(j-1)$, update the current inverse Hessian approximation W_{j-1} over the step s_{j-1}, using any member of the family to obtain, say, W_j^p, and, alternatively, using the B-Update to obtain say W_j^B. Show that the j'th directions $-W_j^p g_j$ and $-W_j^B g_j$ obtained for these two cases are parallel, *and that each is also parallel to* w_{j-1}. Thus $s_j = \gamma w_{j-1}$ for some number $\gamma \neq 0$. Using this result, show that the matrix obtained by updating W_j^p over s_j *using the B-Update* is the same as the matrix obtained by updating W_j^B over s_j *using the B-Update*. This latter observation employs the convenient expression for the B-Update (13) at the root of the family (14), and, in particular, the following fact:

$$\left(I - \frac{s_{j-1} y_{j-1}^T}{y_{j-1}^T s_{j-1}} \right) s_{j-1} = \left(I - \frac{w_{j-1} y_{j-1}^T}{y_{j-1}^T w_{j-1}} \right) (\gamma w_{j-1}) = 0.$$

Then, a straightforward induction establishes both theorems in tandem. For additional detail, see Dixon [1972] and Powell [1972]. A succinct outline is given in Nazareth [1986a], along with an appropriate extension of these two theorems for the inexact line search case.

When applied to a strictly convex quadratic function with $M_1 = I$, Procedure QN/μ satisfies the three properties of Proecdure QN/B given at the beginning of Chapter 1, Section 9. The implication that iterates are identical is just a special case of Dixon's theorem.

The other properties of Procedure QN/B (or equivalently Procedure VM/B) on strictly convex quadratics given in Chapter 1, Section 4.3 ($M_1 = I$) and at the conclusion of Section 10 ($M_1 > 0$) are also true for Procedure QN/μ.

2.1.5   The BFGS Variable-Metric Method

The choice $\mu = 0$, corresponding to the B-Update, is widely regarded to be one of the most effective, and it provides the basis for the BFGS variable-metric method. The conceptual method is embodied in Procedure VM/B of Chapter 2, Section 4.3, and ways to make this mathematical algorithm effective in practice are given in Chapter 5, Section 2.1 and Chapter 6, Section 2.1.2.

2.2   Model-Based QN

The SR1 update is given by

$$M_+ = M + \frac{(y - Ms)(y - Ms)^T}{(y - Ms)^T s} \tag{18}$$

and its inverse $W_+ = M_+^{-1}$ is

$$W_+ = W + \frac{(s - Wy)(s - Wy)^T}{(s - Wy)^T y}. \tag{19}$$

where $W = M^{-1}$.

The update with direct prediction steps, i.e., $\alpha_j = 1 \; \forall \; j$, was proposed in Davidon's original paper [1959], but the variable-metric point of view required that the SR1 update formula be modified to ensure positive definiteness and thus yield an associated metric. This resulted in a non-competitive update, which was relegated to an appendix in Davidon [1959]. For background and a detailed discussion of metric-based SR1 updates, see Smith and Nazareth [1993]. The cycle of rediscovery and renewed enthusiasm for the SR1 update, followed by a 'fall from grace', was subsequently repeated more than once. For contributions, see, for example, Davidon [1968], Fiacco and McCormick [1968], Pearson [1969], Broyden [1967], Brayton and Cullum [1979]. The breakthrough came when Conn, Gould and Toint [1991] showed that the SR1-Update was best suited for use within a model-based method, where positive definiteness is no longer required. See also Khalfan, Byrd and Schnabel [1993].

The SR1-Update has some very attractive properties that are now discussed.

2.2.1 SR1 Characteristics

The update has a very simple and elegant expression (18), and it is *unique*, being the only symmetric rank-one update of a matrix that satisfies the QN relation (see Chapter 1, Section 7.1). Invariance with respect to transformation of variables can be established as in Section 2.1.1.

The update has the *hereditary property* on quadratics under very weak assumptions on line searches, which are not required to be exact, and on the steps over which the update is developed, which are not required to be conjugate. It finds the true Hessian of a strictly convex quadratic in n steps and the minimizing point in $n + 1$ steps (assuming it does not fail or terminate prematurely). This was discussed in Chapter 1, Section 7.1.

The SR1-Update is a member of the B-Family, see Section 2.2.3 below.

The inherent weakness of the SR1-Update in a metric-based setting (not necessarily preserving positive definiteness) can be turned to a strength in a model-based setting, because the update can more accurately model the indefinite Hessians of non-convex functions.

2.2.2 SR1 ∈ Self-Complementary Family

Apply the complementing operation (5) to (18) to obtain

$$W_+ = W + \frac{(s - Wy)(s - Wy)^T}{(s - Wy)^T y}. \tag{20}$$

Observe that (20) is identical to the inverse of M_+ defined by (19). Thus the operations of complementing and inverting produce the *same* update. The SR1-Update is said to be *self-complementary*. In fact, it is a member of a family of self-complementary updates, and other members are referenced in Chapter 6, Section 2.1.1.

2.2.3 SR1 ∈ B-Family

The SR1 is a member of the B-Family of Section 2.1.3. It corresponds to the choice

$$\mu_{SR1} = \frac{b}{b - a}, \quad \nu_{SR1} = \frac{b}{b - c}. \tag{21}$$

Note that μ_{SR1} is not necessarily greater than μ^* defined by (15), so the SR1-Update does not necessarily preserve positive definiteness. The update is singular when

$$\frac{b}{b - a} = \frac{b^2}{b^2 - ac}$$

i.e., $b = c$. Note that this is just when the denominator in the inverse form (19) is zero, i.e., the inverse is infinite. Also, when the denominator in (19) is positive, i.e., $b > c$, then the SR1-Update has hereditary positive definiteness as can be seen directly from (19): $W > 0$ implies that $z^T W_+^{SR1} s > 0$ for any non-zero $z \in \Re^n$. This result can also be obtained from (21) and (16).

By virtue of membership in the B-Family, the SR1-Update has the properties of the family, in particular, invariance under transformations of variables. When used in a metric-based setting, it also satisfies Dixon's and Powell's theorems over any sequence of contiguous steps for which the update remains positive definite.

2.2.4 The SR1 Secant Method

As we have noted, the SR1 update is most appropriately used in a model-based context, in a development that completely parallels the model-based Newton method of Section 1.2. This development was introduced in Chapter 3, Section 3, and we now enlarge on the different aspects of the method.

1. *Factorization*: Since the Hessian approximation M is updated at each step by a rank-one update, we have the efficiency enhancing possibility of updating a factorized representation of M, in particular, an eigendecomposition or a Bunch-Parlett factorization.

   Thus, given the eigendecomposition $M = Q\Lambda Q^T$, where Q is an orthogonal matrix and Λ is diagonal, we have

   $$Q^T M_+ Q = Q^T MQ + \frac{Q^T(y - Ms)(y - Ms)^T Q}{(y - Ms)^T s} = \Lambda + \frac{uu^T}{(y - Ms)^T s} \quad (22)$$

   where $u = Q^T(y - Ms)$. The eigendecomposition of M_+ can be obtained more efficiently by taking advantage of the simpler matrix defined by the right-hand side of (22).

   For a discussion of updating the Bunch-Parlett factorization, see Sorensen [1977].

2. *Ball-Constraint*: Analogous to Section 1.2.2.

3. *Optimal Multiplier*: Analogous to Section 1.2.3. Efficiency enhancements in maintaining a factorization of the update discussed above will translate into efficiency enhancements for solving associated systems of equations involved in finding the optimal multiplier.

4. *Improving Point*: Analogous to Section 1.2.4. Note, however, that a negative curvature direction of the model is not necessarily a negative curvature direction of the function.

- *Line Search Approach*: This approach has not been explored to date and is currently under investigation; see also Smith and Nazareth [1993].

- *Trust Region Approach:*. For details, see, in particular, Conn, Gould and Toint [1991] and Khalfan, Byrd and Schnabel [1993].

3 CG-Related Methods

As we have seen in Chapter 1, the Conjugate Gradient method can be formulated via the minimization of convex quadratic functions and notions of conjugacy, as originally developed by Hestenes and Stiefel [1952]. (For a comprehensive discussion, see Hestenes [1980].) Fletcher and Reeves [1964] then generalized the approach so it could be applied to minimizing an arbitrary smooth function. In essence, a direction is chosen in the subspace spanned by the current negative gradient and the previous step, and in such a way as to parallel the Hestenes-Stiefel direction in the setting of convex quadratics with exact line searches. Various expressions for the linear combination of gradient and step are equivalent in this context (see Chapter 1, Section 4.4). On nonquadratic functions, the Hestenes-Stiefel and Polak-Ribiere choices are identical when line searches are exact. On general functions with inexact line searches, the Hestenes-Stiefel, Polak-Ribiere and Fletcher-Reeves directions are not necessarily parallel. In all three cases, line searches must be sufficiently accurate to ensure a direction of descent.

The CG method can be approached directly within the setting of nonlinear minimization as discussed in Chapter 2, Section 5. This is defined by the direction of steepest descent in the CG-metric, see (2:59). With exact line searches, this direction is the Hestenes-Stiefel (or Polak-Ribiere) direction. When line searches are not exact, a *comprehensive* computational comparison of the relative merits of the the direction (2:59) (or one of its CG variants) and the direction used in Procedure LVM/B of Chapter 2, Section 5, has not been reported in the literature. An effective CG-related algorithm based on the the direction used in Procedure LVM/B has been formulated by Shanno [1978] (see also Chapter 6, Section 3.2), but unfortunately this article does not fully address the issue of comparing the various directions *in their pure forms*, because of the mix of a number of different strategies in the algorithm. Nocedal [1980] provides some pertinent numerical information, and for further discussion, see also Nazareth [1986b].

The foregoing more general derivation of the CG method is simple and elegant, but it still has drawbacks. It depends specifically on the B-Update and exact line searches. In contrast, Newton and quasi-Newton methods each rely on a fundamental unifying idea. Newton methods use a quadratic approximation derived from a Taylor expansion. Quasi-Newton methods rely on a Mean Value Theorem that justifies the QN Relation, coupled with low-rank updates.

Both Newton and quasi-Newton methods have metric-based and model-based versions. A more basic unifying idea is required for CG-related methods as well. This is the motivation for *Successive Affine Reduction* as developed in Nazareth [1986b,c,d] and the idea can be stated very simply as follows: *make estimates of curvature in a Newton or quasi-Newton sense, metric-based or model-based, in an affine subspace defined by the recent history of iterates (gradients and steps)*. When only the most recent gradient and step are used, the affine reduction method is a very close relative of, and under appropriate assumptions, mathematically equivalent to the CG method.

The subject of CG-related methods for minimizing high-dimensional functions is very much in flux and currently an active area of optimization research. Chapter 6, Section 3 will furnish an overview.

The successive affine reduction approach will be considered in some more detail in the subsections below.

3.1 Metric-Based SAR

Let us consider the most basic case, when the affine reduced space is of dimension two and is defined by the current gradient and the previous step, as in the standard CG method.

Let j be the index of a sequence of iterates, and consider a non-zero step s_j from an iterate x_j (with gradient g_j) to x_{j+1} (with gradient g_{j+1}). Thus $s_j = x_{j+1} - x_j$, and, as usual, $y_j = g_{j+1} - g_j$.

Assume s_j and g_{j+1} are linearly independent and orthonormalize these two vectors to obtain the $n \times 2$ matrix $Q_j = [\hat{s}_j, \hat{g}_{j+1}]$. Make the transformation (affine reduction) at x_{j+1} as follows:

$$x = x_{j+1} + Q_j \tilde{x} \tag{23}$$

where $\tilde{x} \in \Re^2$.

Suppose g and H denote the gradient vector and Hessian matrix at x, and $\tilde{g}$ and $\tilde{H}$ the corresponding transformed quantities. Then by the chain rule $\tilde{g} = Q_j^T g$ and $\tilde{H} = Q_j^T H Q_j$. The step s_j and gradient-change y_j transform as follows: $s_j = Q_j \tilde{s}_j$ and $\tilde{y}_j = Q_j^T y_j$. Also, $\tilde{s}_j = \begin{bmatrix} 1 \\ 0 \end{bmatrix} \|s_j\|_2$.

3.1.1 Affine Reduced Modified-Newton

The reduced 2×2 Hessian matrix $\tilde{H}_{j+1}$ at $\tilde{x}_{j+1}$ is estimated by finite differences, using just a few function evaluations, and it can be modified, if necessary, in order to obtain a positive definite matrix $\tilde{H}'_{j+1}$. Define a corresponding full $n \times n$ positive definite Hessian approximation by

$$H'_{j+1} = [Q_j | Q_j^*] \begin{bmatrix} \tilde{H}'_{j+1} & 0 \\ 0 & I \end{bmatrix} \begin{bmatrix} Q_j^T \\ Q_j^{*T} \end{bmatrix},$$

where Q_j^* has orthonormal columns that span the orthogonal complement of Q_j. Since $Q_j^{*T} g_{j+1} = 0$, the matrix Q_j^* is not needed in order to obtain the resulting modified-Newton direction, which is given by $d_{j+1} = -Q_j [\tilde{H}'_{j+1}]^{-1} Q_j^T g_{j+1}$.

As in Chapter 2, we can summarize the essential details of the resulting conceptual procedure as follows:

Procedure SAR/MN

Given an initial point x_1, gradient g_1 and direction $d_1 = -g_1$:

for $j = 1, 2, \ldots,$

   $\alpha_j^* = \mathrm{argmin}_{\alpha \geq 0} f(x_j + \alpha d_j)$.

   $x_{j+1} = x_j + \alpha_j^* d_j$. Evaluate g_{j+1} and $s_j = x_{j+1} - x_j$.

   Othonormalize s_j and g_{j+1}, and define $Q_j = [\hat{s}_j, \hat{g}_{j+1}]$.

   Form the following quantities:
$$
\begin{aligned}
\alpha &= [f(x_{j+1} + h\hat{s}_j) - 2f(x_{j+1}) + f(x_{j+1} - h\hat{s}_j)]/h^2 \\
\beta &= [f(x_{j+1} + h(\hat{s}_j + \hat{g}_{j+1})) - f(x_{j+1} + h\hat{g}_{j+1}) - f(x_{j+1} + h\hat{s}_j) + f(x_{j+1})]/h^2 \\
\gamma &= [f(x_{j+1} + h\hat{g}_{j+1}) - 2f(x_{j+1}) + f(x_{j+1} - h\hat{g}_{j+1})]/h^2
\end{aligned}
$$

where h is a small positive step length. Define the matrix $\tilde{H}_{j+1} = \begin{bmatrix} \alpha & \beta \\ \beta & \gamma \end{bmatrix}$ and compute its eigenvalues, say $\lambda^{\pm}$. Let $\lambda = \min\{\lambda^+, \lambda^-\}$. If $\lambda \leq \delta$, where δ is a small positive constant, then modify $\tilde{H}_{j+1}$ to $\tilde{H}'_{j+1} > 0$ (see Section 1.2). Otherwise, $\tilde{H}'_{j+1} = \tilde{H}_{j+1}$.

   Set $r = Q_j^T g_{j+1}$. Solve $\tilde{H}'_{j+1} v = r$ and form $d_{j+1} = -Q_j v$.

end

When j is a multiple of n, or when s_j and g_{j+1} are close to linear dependency, restart the foregoing procedure.

Procedure SAR/MN is primarily of interest when the evaluation of the function is much cheaper than the evaluation of its gradient. For further discussion and some illustrative numerical results, see Nazareth [1992].

When line searches are exact and the function is a strictly convex quadratic, Procedure SAR/MN is equivalent to the standard Hestenes-Stiefel CG method, see Nazareth [1986d].

3.1.2 Affine Reduced Variable Metric

Now we consider the case when the Hessian approximation is developed by a variable metric method.

Given an approximation M_j to the Hessian matrix, an updated approximation M_{j+1} will be developed as follows:

1. $\bar{M}_j = Q_j^T M_j Q_j$ defines a reduced Hessian approximation.

2. Update $\bar{M}_j$ to $\bar{M}_{j+1}$ by the B-Update over the reduced step and gradient change pair $(\tilde{s}_j, \tilde{y}_j)$.

3. Let

$$
M_{j+1} = [Q_j | Q_j^*] \begin{bmatrix} \tilde{M}_{j+1} & 0 \\ 0 & I \end{bmatrix} \begin{bmatrix} Q_j^T \\ Q_j^{*T} \end{bmatrix} = Q_j \tilde{M}_{j+1} Q_j^T + (I - Q_j Q_j^T)
$$

where Q_j^* has orthonormal columns that span the orthogonal complement of Q_j. The matrix M_{j+1} is *implicitly* defined by Q_j and $\tilde{M}_{j+1}$.

Note that M_j will itself be implicitly defined by two matrices Q_{j-1} and $\tilde{M}_j$ from the previous iteration, and the search direction d_{j+1} is given by $d_{j+1} = -Q_j[\tilde{M}_{j+1}]^{-1} Q_j^T g_{j+1}$.

We can summarize the essential details of the resulting conceptual procedure as follows:

Procedure SAR/VM:

Given an initial point x_1 with gradient g_1:

for $j = 1, 2, \ldots$,

$d_j = -Q_{j-1} \tilde{M}_j Q_{j-1}^T g_j$. (Initially, $d_1 = -g_1$.)

$\alpha_j^* = \operatorname{argmin}_{\alpha \geq 0} f(x_j + \alpha d_j)$.

$x_{j+1} = x_j + \alpha_j^* d_j$. Evaluate g_{j+1}, $s_j = x_{j+1} - x_j$ and $y_j = g_{j+1} - g_j$.

Orthonormalize s_j and g_{j+1} to obtain $Q_j = [\hat{s}_j, \hat{g}_{j+1}]$.

Revise the approximation to the reduced Hessian in the affine subspace defined by Q_j as follows:

$P_j = Q_j^T Q_{j-1}$. (Initially $P_1 = I$.)

$\tilde{s}_j = \begin{bmatrix} 1 \\ 0 \end{bmatrix} \|s_j\|$; $\tilde{y}_j = Q_j^T y_j$. (Note that $\tilde{s}_j \tilde{y}_j > 0$.)

$$\bar{M}_j = I + P_j(\tilde{M}_j - I)P_j^T. \text{ (Initially, } \tilde{M}_j = I.)$$
$$\tilde{M}_{j+1} = \bar{M}_j - \frac{\bar{M}_j \tilde{s}_j \tilde{s}_j^T \bar{M}_j}{\tilde{s}_j^T \bar{M}_j \tilde{s}_j} + \frac{\tilde{y}_j \tilde{y}_j^T}{\tilde{y}_j^T \tilde{s}_j}.$$

end

When j is a multiple of n or when s_j and g_{j+1} are close to linear dependency, *restart* the foregoing procedure.

For further discussion and some numerical results on problems of modest dimension, see Nazareth [1986c]. In particular, Procedure SAR/VM can be shown to be equivalent to the Hestenes-Stiefel CG algorithm on arbitrary smooth functions when line searches are exact. When line searches are not exact, it is a refinement of Perry's algorithm [1976] that always produces a direction of descent under the standard line-search restriction $y_j^T s_j > 0$.

3.1.3 Discussion

The use of a two-dimensional affine reduced space is the simplest choice, and as we have seen, the SAR method then represents a natural generalization of the more standard CG methods. Many other variants and extensions, for example, to subspaces of higher dimension, are possible. See Nazareth [1986b,c,d] for further detail. There are also interesting connections with the methods of Khoda and Storey [1990], Khoda, Liu and Storey [1992].

3.2 Model-Based SAR

Affine Reduced Newton and Affine Reduced Secant methods remain to be explored.

4 Cauchy Methods

Steepest descent originates with Cauchy [1829], and we shall view any method based on steepest descent in the metric defined by a *diagonal* matrix, either fixed or variable, as a Cauchy method. When the elements of the diagonal matrix are restricted to be positive numbers, then the method is metric-based, otherwise it is model-based. Note also that a variable diagonal matrix is potentially a rank-n change.

4.1 Metric-Based Cauchy

See Chapter 2, Section 3. The updating of diagonal matrices is an interesting open problem, see Gilbert and Lemarechal [1989] and also Gill and Murray [1979].

4.2 Model-Based Cauchy

See Chapter 3, Section 5. This approach remains to be fully explored.

CHAPTER 5
CONVERGENT IMPLEMENTABLE ALGORITHMS

1 Introduction

In the previous chapter, we described the Newton/Cauchy framework of uncon-strained optimization methods. A *method*, in turn, gives rise to *mathematical algorithms*, i.e., precise computational prescriptions whose convergence must be formally addressed. In a terminology popularized by Polak [1971], math-ematical algorithms are further classified into two categories: *conceptual* and *implementable*. The former are permitted to include operations that are not necessarily finite. For example, the procedures of Chapter 2 are conceptual mathematical algorithms. They involve an exact line search on an arbitrary smooth function to obtain the steplength α_j^*, which cannot be found in a finite number of operations. Implementable algorithms, on the other hand, must only contain finite operations. For example, the procedures in Chapter 1 are imple-mentable mathematical algorithms, because all operations, including finding α_j^* on a convex quadratic function, can be done in a finite number of steps.

We now turn to the derivation of implementable algorithms from conceptual algorithms developed earlier, and, in particular, we will describe some simple but extremely powerful results that can be used to demonstrate global conver-gence (in an appropriate sense) of mathematical algorithms, both conceptual and implementable.

An optimization method must eventually be encoded into a numerically sta-ble computer implementation and software that puts it to practical use. We discuss the important idea of *hierarchical implementation*: the successive stages in the development of an optimization method, as it evolves from a concep-tual/implementable mathematical algorithm, to a stable numerical algorithm, and eventually to sound mathematical software.

2 Convergence Conditions

Assume that $f : \Re^n \to \Re^1$ is a continuously differentiable function that is bounded below in $\Re^n$ and whose gradient mapping $g : \Re^n \to \Re^n$ is Lipschitz continuous, i.e., there exists a constant $L > 0$ such that

$$\|g(x) - g(\bar{x})\|_2 \leq L\|x - \bar{x}\|_2 \ \ \forall \ x, \bar{x} \in \Re^n. \tag{1}$$

Consider any iteration of the form

$$x_{j+1} = x_j + \alpha_j d_j, \tag{2}$$

where α_j denotes a positive steplength and d_j is a *direction of descent*, i.e., $d_j^T g_j < 0$, where g_j denotes the gradient vector at x_j.

Denote the angle between the negative gradient and d_j by θ_j, which is thus given by

$$\cos\theta_j = \frac{-g_j^T d_j}{\|g_j\|_2 \|d_j\|_2} > 0. \tag{3}$$

Previously α_j was chosen by an exact line search, i.e., $\alpha_j = \alpha_j^*$. Now we seek more relaxed conditions that nevertheless ensure sufficient progress *both in terms of function reduction and progress (distance) from the current point*.[1]

2.1 Armijo-Goldstein-Wolfe Conditions on a Line Search

The steplength $\alpha_j \geq 0$ along the descent search direction d_j, is chosen to satisfy the following two conditions:

$$f(x_{j+1}) \leq f(x_j) + \epsilon \, \alpha_j g_j^T d_j, \ \epsilon \in (0,1). \tag{4}$$

and

$$g_{j+1}^T d_j \geq \tau \, g_j^T d_j, \ \tau \in (\epsilon, 1). \tag{5}$$

These are called the *Wolfe Conditions* (Wolfe [1969, 1971]). See also Armijo [1966], Goldstein [1967].

We first show that it is always possible to find a point that satisfies the Wolfe conditions as follows:

From the Taylor expansion,

$$f(x_j + \alpha d_j) = f(x_j) + \alpha g_j^T d_j + R(\alpha),$$

where the remainder term $R(\alpha) = O(\alpha^2)$. Since $g_j^T d_j < 0$ and $\epsilon \in (0,1)$, clearly for all α sufficiently small,

$$\rho(\alpha) \equiv f(x_j + \alpha d_j) < f(x_j) + (\epsilon g_j^T d_j)\alpha.$$

Since the right hand side defines a line with negative slope, and by assumption f is bounded below, the line must intersect $\rho(\alpha)$ at least once. Let $\hat\alpha$ be the first intersection, i.e., the smallest value of α such that

$$\rho(\hat\alpha) \equiv f(x_j + \hat\alpha d_j) = f(x_j) + (\epsilon g_j^T d_j)\hat\alpha. \tag{6}$$

Let $\hat x = x_j + \hat\alpha d_j$. Then clearly the Wolfe condition (4) is satisfied for all $x \in (x_j, \hat x)$.

---

[1] It is insufficient to simply require that $f(x_{j+1}) < f(x_j)$ and that d_j be a direction of descent. For counterexamples, see Dennis and Schnabel [1983].

Next, using a standard mean value theorem for unidimensional functions, there exists an $\bar{\alpha} \in (0, \hat{\alpha})$ such that

$$
\begin{aligned}
f(\hat{x}) - f(x_j) &= g(x_j + \bar{\alpha}d_j)^T(\hat{x} - x_j) \\
&= g(x_j + \bar{\alpha}d_j)^T \hat{\alpha}d_j.
\end{aligned}
$$

Thus

$$
\frac{f(\hat{x}) - f(x_j)}{\hat{\alpha}} = g(x_j + \bar{\alpha}d_j)^T d_j.
$$

Then using (6),

$$
g(x_j + \bar{\alpha}d_j)^T d_j = \epsilon g_j^T d_j > \tau g_j^T d_j,
$$

since $\tau > \epsilon$ and $g_j^T d_j < 0$. Therefore the point $\bar{x} = x_j + \bar{\alpha}d_j$ also satisfies the second Wolfe condition (5). □.

Furthermore, by continuity arguments, there are a *range of points around* $\bar{\alpha}$ with corresponding values $\bar{x}$ that satisfy the two Wolfe conditions.

For typical values of the parameters, for example, $\epsilon = 10^{-4}$ and $\tau = 0.9$, the Wolfe conditions are very weak. The second Wolfe condition (5) is often called the one-sided Wolfe condition to distinguish it from the stronger *two-sided* condition:

$$
\frac{|g_{j+1}^T d_j|}{|g_j^T d_j|} \leq \tau_2, \ \tau_2 \in [0, 1). \tag{7}
$$

By taking $\tau_2 = 0$, the condition (7) can be used to enforce an exact line search in a conceptual mathematical algorithm. It is easy to verify that the two-sided condition implies the one-sided condition, and also that the Wolfe conditions (4)-(5) imply that $y_j^T s_j > 0$, where $s_j = x_{j+1} - x_j$ and $y_j = g_{j+1} - g_j$.

An effective line search procedure that implements the Wolfe conditions is given in Moré and Thuente [1990].

2.2 Wolfe ⇒ Zoutendijk

When the iteration defined by (2) satisfies the Wolfe conditions (4) - (5) then

$$
\sum_{k=1}^{\infty} \cos^2 \theta_j \|g_j\|_2^2 < 0, \tag{8}
$$

where θ_j is defined by (3).

To establish this simple but powerful result, we proceed as follows:

$$
\begin{aligned}
|(g_{j+1} - g_j)^T d_j| &\leq \|g_{j+1} - g_j\|_2 \|d_j\|_2 \\
&\leq L\|x_{j+1} - x_j\|_2 \|d_j\|_2 \ \text{using (1)} \\
&= L\alpha_j \|d\|_2^2. \tag{9}
\end{aligned}
$$

Since (5) implies that $(g_{j+1} - g_j)^T d_j \geq 0$,

$$(g_{j+1} - g_j)^T d_j \leq L\alpha_j \|d_j\|_2^2.$$

Again from (5),

$$(g_{j+1} - g_j)^T d_j \geq (\tau - 1)g_j^T d_j.$$

Thus from (9),

$$L\alpha_j \|d\|_2^2 \geq (\tau - 1)g_j^T d_j.$$

Thus

$$\alpha_j \geq \frac{(\tau - 1)}{L} \frac{g_j^T d_j}{\|d_j\|_2^2}.$$

From (4),

$$f(x_{j+1}) \leq f(x_j) + \frac{\epsilon(\tau - 1)}{L} \frac{(g_j^T d_j)^2}{\|d_j\|^2}.$$

Then using (3),

$$f(x_{j+1}) \leq f(x_j) + \frac{\epsilon(\tau - 1)}{L} \cos^2\theta_j \|g_j\|_2^2.$$

When we sum the foregoing expression over all value of k and recall that the function is assumed to be bounded below, we immediately obtain

$$\frac{-\epsilon(\tau - 1)}{L} \sum_{k=1}^{\infty} \cos^2\theta_j \|g_j\|_2^2 < \infty.$$

Thus, since $\tau < 1$,

$$\sum_{k=1}^{\infty} \cos^2\theta_j \|g_j\|_2^2 < \infty. \quad \Box$$

This is the *Zoutendijk Condition* (Zoutendijk [1970]).

2.3 Zoutendijk $\Rightarrow$ Convergence

As soon as we assume that $\cos\theta_j$ is bounded away from zero, i.e., the angle θ_j is bounded away from 90 degrees, then the Zoutendijk condition immediately implies that

$$\lim_{k \to \infty} \|g_j\|_2 = 0,$$

i.e., the algorithm in the limit generates iterates (2) that are attracted to a stationary point. Note that the Wolfe and Zoutendijk conditions are not sufficient to ensure that the the iterates themselves converge.

Most of the methods discussed in previous chapters that use a line search produce a descent direction by solving a linear system of the form

$$\bar{H}_j d_j = -g_j, \tag{10}$$

where $\bar{H}_j$ is a symmetric positive definite matrix. Suppose $\bar{H}_j$ has a *bounded* condition number, i.e., $\frac{\sigma_1}{\sigma_n} \leq \Delta_H$, where $\sigma_1 > 0$ and $\sigma_n > 0$ denote the largest and smallest eigenvalues of $\bar{H}_j$ and Δ_H is a given (generally large) positive number. Then the angle θ_j is bounded away from 90 degrees. This can be seen as follows:

$$\cos\theta_j = -\frac{g_j^T(-\bar{H}_j g_j)}{\|g_j\|_2 \|\bar{H}_j g_j\|_2} \geq \frac{g_j^T \bar{H}_j g_j}{\|g_j\|_2 \|\bar{H}_j\|_2 \|g_j\|_2} \geq \frac{u_j^T \bar{H}_j u_j}{\|\bar{H}_j\|_2}. \tag{11}$$

where $u_j = g_j / \|g_j\|_2$ is a vector of unit length. Then the last term of (11) is bounded below by $\frac{\sigma_n}{\sigma_1} \geq \frac{1}{\Delta_H}$. $\square$

The foregoing results are broadly applicable to implementable algorithms derived from the methods of Chapter 4. In metric-based methods, $\bar{H}_j$ is usually maintained in a factored form $\bar{R}_j^T \bar{R}_j$, and the condition number can be bounded by controlling that of the reconditioner $\bar{R}_j$. The latter operation is often facilitated by maintaining $\bar{R}_j$ in triangular form. In model-based methods, $\bar{H}_j$ is of the form $(H_j + \lambda_j I) > 0$ or $(M_j + \lambda_j I) > 0$, and its condition number can be bounded via λ_j.

We leave the exploration of particular instances to the reader. Here we have simply sought to open a window on the rich theory of *global convergence*. Other important aspects of convergence analysis are *local rate of convergence* and *global efficiency* (or *complexity*) of unconstrained optimization algorithms. A detailed discussion of these three topics is not within the scope of this monograph. For classical convergence results see Goldstein [1967], Ortega and Rheinboldt [1970] and Dennis and Schnabel [1983]. For three outstanding surveys articles, see Dennis and Moré [1977], Moré and Sorensen [1982], and Nocedal [1991]. For the convergence of trust-region algorithms, see Moré [1983]. And, for complexity results, see Luenberger [1984] (convex quadratics) and Nemirovsky and Yudin [1983], Nesterov [1983] (convex functions).

3 Hierarchical Implementation of Optimization Methods

In the early stages of formulation of an optimization method into a mathematical algorithm, it is useful to be able to implement it in a high-level language that uses the vernacular of computational mathematics. Such a language serves as a medium for communicating algorithmic ideas precisely. It permits coding and subsequent modification to be carried out with relative ease, even when this results in a sacrifice of efficiency with regard to computer time, computer storage and numerical stability, and, in consequence, permits only fairly artificial problems or only a subclass of realistic problems to be solved. Later it becomes necessary to attend to the practicalities of computation, namely, finite-precision arithmetic and the limitations of computer resources. Thus questions of numerical stability and efficiency must be addressed and they usually imply substantial reformulation of data structures and calculations, leading to an implementation which is capable of solving problems more representative of those encountered in practice. Such an implementation provides more realistic evidence about algorithm performance and also addresses the research needs of the experienced user. (By experienced user we mean one familiar with both the problem being solved and the algorithm used to solve it.) Finally, we come to the everyday user who is not expected to be familiar with the details of the algorithm. Meeting his or her requirements requires another giant step in elaboration and refinement in order to obtain a production code or item of mathematical software which has a convenient user interface, is transportable, and solves problems automatically with a minimum of user intervention.

It is thus useful to think in terms of *a hierarchy of implementations* of an optimization method, and convenient to distinguish three levels in the hierarchy corresponding to the initial, intermediate and final stages of implementation as outlined above. These hierarchical levels can also be viewed as representing the stages of evolution from formulation and implementation of a *mathematical algorithm*, through reformulation and implementation of a viable *numerical algorithm*, to final implementation of good *mathematical software*.

3.1 Mathematical Algorithms and Level-1 Implementations

At this root level, we remain, in a sense, within the domain of the algorithm inventor and computational mathematician. We are concerned here with algorithmic theory that relates a relatively small number of individual mathematical algorithms to one another, thus providing a coherent overall structure, and with the formulation and analysis of these individual algorithms. Issues and analysis of global convergence, local rate of convergence and complexity play a central role. At this level, one seeks to make the task of implementing a particular mathematical algorithm (or testing a conjecture about some property of an algorithm) as painless as possible, so as to obtain some initial computational experience. This experience often results in new insights and thus helps in laying out the basic features of an algorithm.

An interactive high-level language which uses the vernacular of computational mathematics provides a convenient "sketch-pad" for working out algorithmic ideas or verifying conjectures on a computer. Ideally such a language should provide a wide array of mathematical operators and data types and should also be easy to extend. The APL language was a pioneering effort of this type. It can be used interactively, has convenient facilities for defining new operators and makes the full facilities of the language available for investigating relationships between program variables at a breakpoint. A disadvantage is that the compact notation often makes programs difficult to read or decipher.

Speakeasy has the power of APL, but is much more user-friendly. It makes possible very readable programs and provides an interface and language extension mechanism that permits independently developed Fortran programs to be conveniently incorporated into the language. When APL and Speakeasy are used in interpretive mode, there is often a price to be paid in terms of program efficiency.

More recent and very suitable languages for the purpose include Gauss, Matlab, Maple and Mathematica. (For example, the procedures of Chapters 1 and 2 could be implemented very quickly in Matlab using its matrix operators and an unidimensional minimizer provided with its optimization toolbox.) Also, there is obviously no reason why a modern dialect of Fortran should not be used, particularly when it is enhanced by a collection of modules that makes available a wide range of mathematical operators. We shall refer to experimental implementations developed in an extensible high-level language as level-1 implementations.

A level-1 routine is useful not only to give a precise definition to an algorithmic idea, but also to serve as a medium for communicating the ideas. In particular, it provides a useful teaching tool. Often the main reason for developing a level-1 implementation is experimentation so rigorous standards of implementation need not be applied. A 'quick and dirty' program involving unrealistic data management and a blasé attitude to potential numerical difficulties is often initially acceptable. Although one sometimes runs the risk of obtaining misleading computational information because of numerical error, a good algorithm should not be unduly victimized by finite-precision arithmetic. In addition to revealing conceptual difficulties, experimentation at this level can also reveal potential numerical difficulties which can then be addressed when developing a higher-level code. The aims are often geared to testing *viability* of an algorithm, i.e., the implementation may not be good enough to measure either efficiency or reliability, which are important testing criteria at subsequent levels.

A level-1 implementation will often incorporate just the *key features of an algorithm*. Often the problems which are addressed may be artificial or a subset of realistic problems, for example, smooth convex functions in nonlinear minimization. Convenience of problem specification may be emphasized. And often a level-1 implementation will be designed simply to solve a *particular* problem, i.e., to 'get the job done'. Level-1 routines are generally programs with a short shelf life, which are intended to provide initial computational feedback or to furnish the solution of a particular problem, and they will then often be discarded or modified beyond recognition.

3.2 Numerical Algorithms and Level-2 Implementations

A high-level language, no matter how convenient, does not circumvent (although it may postpone) the painstaking task involved in implementing a routine that is both numerically sound and efficient in its use of computer resources. At level-2, we are concerned with making a viable algorithmic idea workable within the setting of finite-precision arithmetic, the limitations of computer resources and a problem space that is more representative of problems that arise in practice. We are now, so to speak, firmly within the province of the computational scientist and numerical analyst. For example, Dahlquist and Bjorck [1974] define a numerical algorithm as a complete description of well-defined operations whereby each permissible input data vector is transformed into an output data vector. By 'operations' is meant the atomic and logical operations which a computer can perform, together with references to previously defined numerical algorithms. Round-off error analysis is a measure of the intellectual difficulties associated with algorithm reformulation and implementation at this level, see Wilkinson [1965]. Since a particular mathematical algorithm can be reformulated in a variety of ways, each with different numerical properties, we are in a larger 'algorithmic space'. For example, e^x can be computed by the series $\sum_{i=1}^{N}(x^i/i)$ for some suitable N. However, if x is negative, cancellation error cn be avoided by reformulating this calculation as $1/\sum_{i=1}^{N}\{(-x)^i/i\}$. Algorithms at level-2 sometimes bear little resemblance to one another even when they are based on the same mathematical formulae, because the reformulation can be at a very basic level of detail.

Since one is concerned here with wider distribution of programs than at level-1 and with programs that have a longer 'shelf-life', it is more likely that Fortran, Algol, C, Pascal or some similar language would be used (or a pseudo-code based on one of these languages). Since the variety of operators in these languages is quite limited, a collection of modules becomes especially useful here (modules cn be thought of as a form of language extension). A modular collection also provides a 'sketch-pad' and a teaching tool at a more sophisticated level of implementation than level-1. Important criteria for both modules and level-2 implementations are: *flexibility, ease of modification, generality, readability*. As already mentioned, attention should be paid both to numerical aspects and to data management, but relatively simple strategies (for example, very simple convergence criteria) may still he employed. The aim is to develop readable and modifiable implementations that can be employed by an experienced user to solve research problems and used for algorithm experimentation and development in a realistic problem setting. These two aims, of course, go well together. The goal of readability and modifiability for modules is especially important because it is unlikely that their design will be general enough for them to be used in simply a 'plug-in' mode. Some changes will quite often have to be made to suit the particular implementation at hand.

As regards problem specification, we are concerned here with problems that are representative of those that arise in practice. For nonlinear function minimization, one might wish to consider situations where different sorts of information are available, for example, problems in which only function values are available. Testing would emphasize *efficiency* and *stability*, in order to investigate whether the algorithm is really effective. Effectiveness might be measured in terms of the number of function evaluations needed to find the optimal solution and the number of failures. There would also be the need for probing and tracing within the code as in level-1.

Examples of level-2 implementations are the optimization routines available in the A.E.R.E. Harwell library. From them, Hillstrom [1976a,b] derived both an in-house modular collection and a set of test problems at Argonne National Laboratory. The work described in Nazareth [1977b] represents an in-house experiment in organizing a collection of level-2 implementation aids that built on the modules and test problems of Hillstrom [1976a,b]. This level-2 collection was used, in turn, to implement a set of algorithms for unconstrained minimization and nonlinear least squares proposed in Nazareth [1975] and [1976b,c].

The foregoing efforts and interaction with Argonne National Laboratory motivated the more ambitious effort into developing level-2 modules of Dennis and Schnabel [1983]. They also provided the starting point for development of the widely available test-bed described in Moré, Garbow and Hillstrom [1981].

3.3   Mathematical Software and Level-3 Implementations

Finally we come to the development of user-oriented production codes or mathematical software. This falls within the province of the software engineer and requires yet another giant step in elaboration and refinement in order to arrive at a complete (perhaps even definitive) expression of an algorithm in a particular language (usually Fortran) in a form capable of execution on a range of computers. It may incorporate complex adaptive strategies and tactics in order to further enhance efficiency. Usually such software is expected to meet certain standards of quality and at the very least it should be user-friendly and reliable. Just as the need for fine attention to detail in convergence, complexity and error analysis typify the inherent difficulties addressed at levels 1 and 2, here one might characterize the difficulties by a term such as software *systems analysis* or perhaps, more appropriately, *systems synthesis*. This also requires careful attention to detail although this is less mathematical in nature, and careful coordination of many disparate components is necessary in order to synthesize from them a workable piece of software. Obviously this task also requires an appreciation of the insights which convergence analysis and error analysis can provide and the principles that guide algorithm formulation, such as invariance under transformations of variables.

Fortran being the accepted language of scientific computation, it is the most common choice of implementation choice of implementation language. There is now an extensive literature on the development of quality mathematical software, see for example, Smith et al [1974], Rice [1971] for pioneering work. Quality mathematical software is expected to be *reliable, robust, efficient, structured, well-documented, valid* and *transportable*.

It is likely that in time the word 'quality' or the phrase 'reliable and robust' will be automatically assumed when we talk about mathematical software. After all, no one now talks about reliable and robust harware. It is just expected to have these qualities. It is also likely that the terms 'package' and 'library' will increasingly be associated only with software at this level. As noted, the softwre should be well-structured and modular, but here 'module' is used more in the sense of an efficient subroutine in a structured program, designed within the context of the overall implementation or package. It is much more likely that such modules would be used in a plug-in mode. Often the design of level-3 software is too delicate for causal tampering and sometimes such software is simply not designed to be modifiable.

Obviously, level-3 software should be able to solve problems as they arise in practice, and it is common for a package to include routines to check user-supplied information or to aid the process of specifying problems in a standard format. In the testing process increased emphasis is placed on reliability and robustness, and in the evaluation process routines are often treated as black boxes to be evaluated as one would any consumer product (see, for example, the performance profile approach of Lyness [1979]).

Good examples of level-3 unconstrained minimization software can be found in the Minpack-1 collection, see Moré, Garbow and Hillstrom [1980], [1981], and the Numerical Algorithms Group (NAG) library, see Gill, Murray, Pickens and Wright [1979].

3.4 Discussion

Note that distinctions between the three levels of implementation are not governed by the size of problems addressed. A user-oriented, high-quality code for solving problems in a few variables would be considered a level-3 code.

There is obviously nothing sacrosanct about identifying three as opposed to, say, four levels in the hierarchy of implementation and obviously no clear-cut distinctions between them. Level-1 and 2 routines can and should be used to solve practical problems, and a level-3 implementation can be used to study the encoded algorithm and, by replacing part of the code, to develop and experiment with related algorthms. In general, however, one can say that while higher-level routines, in particular, level-3 routines are useful at a lower level, lower-level routines, in particular level-1, may not be of *direct* use at a higher-level. However, a central theme of this discussion is that a lower-level implementation provides a *staging-ground* for the development of an implementation of an optimization method at a higher level. In particular, *validation* of higher-level routines can be carried out using the results obtained from lower-level routines.

Another theme is that by drawing distinctions between levels of implementation one can usually clarify both the intent of a code and the meaning of terms like module, testing criteria, usability and so on. They are usually determined by the level of implementation. For example, implementing a gradient method with line searches performed by bisection on the directional derivative can be done in 'about 15 minutes' using Matlab. But a good level-3 implementation of steepest-descent would require careful craftsmanship and be much more time consuming. For another example, consider the optimization 'recipes' in Press, Vetterling, Teukolsky and Flannery [1992]. Judged by level-1 criteria, they represent a valuable contribution, but the implementations are very crude when judged by the criteria of levels 2 and 3.

An important added benefit of making explicit distinctions between levels of implementation is that they help to reduce the 'intimidation factor', whereby optimization researchers are reluctant to part with preliminary or experimental software for fear that unfair criticisms will be made of their codes. In software development, the cardinal principle should be that 'the proof of the pudding is in the eating', but it helps to be able to be more specific about what pudding is being served.

With the increasing availability of very high-level languages, and with scientists and engineers now being drawn increasingly into computational mathematics and algorithmics and making important contributions, it is likely that *level-1 implementation will become the center of action*, providing an arena for new algorithmic ideas and an environment for solving practical problems. The latter often have a great deal of special structure that must be taken into consideration in order to make them computationally tractable, and this characteristic makes them less amenable to solution by general purpose level 3 software. To solve such problems often requires the development of tailor-made level-1 implementations. Levels 2 and 3 are more the province of skilled specialists. The level-3 implementations that they eventually develop then percolate back as new operators for use within *high-level languages* at level 1, for example, Speakeasy, Matlab or Mathematics, i.e, hierarchical implementation has a feedback loop that eventually enriches the level-1 environment.

4 Notes

Section 2: The discussion is derived from Dennis and Schnabel [1983] and Nocedal [1991].

Section 3: The discussion is derived from Nazareth [1985]. See also Trefethen [1992].

CHAPTER 6
UNCONSTRAINED OPTIMIZATION TECHNOLOGY

In this monograph, our focus has been the key *methods* of unconstrained nonlinear optimization and the illuminating relationships between them. We discussed the resulting so-called *Newton/Cauchy framework* and described some of the *basic principles* involved in hierarchical implementation, i.e., the process of transforming a conceptual method into a globally convergent implementable mathematical algorithm, a stable numerical algorithm and, eventually, an item of quality mathematical software.

Our treatment has *not* done justice to to the rich array of mathematical and numerical *algorithms* developed in the field of unconstrained nonlinear optimization, nor to the delightful rate of convergence analysis, numerical analysis and implementation techniques that characterize the subject. We can only provide the reader with a *structured guide*[1] to this extensive Newton/Cauchy 'technology', in the present short concluding chapter.

Our overview is far from comprehensive, and, in the interests of brevity, many important contributions are not referenced. Within each category below, we simply cite a few key references, sometimes enlarged through a brief comment. Their bibliographies can, in turn, be used to trace other relevant articles.

1 Newton-Technology

1.1 Variants

When H is indefinite, compute a direction of negative curvature using either the eigendecomposition or the Bunch-Parlett factorization [1971]. Use this direction, possibly in combination with the negative gradient direction, to define a suitable direction of descent, and then employ a line search to obtain an improving point. For variants, see Fletcher and Freeman [1977], Goldfarb [1980], Moré and Sorensen [1982], Sorensen [1982a]. The approach has not really proved itself in practice, see Dennis and Schnabel [1983].

1.2 Finite Differences

Elements of the Hessian matrix can be obtained using differences of gradients or function values. Enhancements are possible when the Hessian matrix is sparse by means of a careful choice of directions along which differences are taken. The approach makes extensive use of graph-theory. See, for example, Coleman and Moré [1982], Powell and Toint [1979].

---

[1] 'road-map' or 'taxonomy'

1.3 Automatic Differentiation

Although no panacea, the ability to derive computer programs that compute first and second derivatives from user-supplied programs that compute only function values has caused a resurgence in the use of Newton's method. Automatic differentiation is obviously also relevant to quasi-Newton and CG-related algorithms. For details, see, for example, Griewank [1989], Gilbert [1992].

1.4 Truncated Newton

The symmetric system of linear equations $Hd = -g$ that defines the search direction d in Newton's method can be computed inexactly, using a version of the linear conjugate gradient algorithm. Furthermore, the information needed by the latter can be estimated using gradient differences along appropriate directions, thus circumventing an explicit Hessian matrix-vector product. This is especially useful in large-scale applications. For details, see, for example, Dembo, Eisenstat and Steihaug[1982], Nash [1985].

1.5 Other Models

For accelerations of Newton's method based on other models, see, for example, Schnabel and Chow [1991: tensor models], Schnabel [1993: tensor models], Nazareth and Ariyawansa [1989; conic models].

2 QN-Technology

2.1 Variants

2.1.1 Alternative Updates

The subject of QN updating is delightfully interesting from a theoretical standpoint, and there is an enormous literature on quasi-Newton updates, derived from the freedom of the parameters corresponding to the elements of Λ in expression (1:50), alternative choices for the vectors u and v in metric-based methods (see (2:33)), the use of rank-2 updates in model-based methods (see (3:33)), and so on. These derivations employ a wide variety of variational principles and the complementing operation (4:5), in particular, in order to obtain self-complementary updates[2]. From a purely practical standpoint, it is currently unclear whether the relatively small gains in computational performance vis-a-vis the SR1 and B-Updates justifies the considerable expenditure of effort involved in inventing, analysing and testing alternative QN Updates.

---

[2]It is not difficult to show that there exists a parameterized *family* of self-complementary updates.

Some relevant references are as follows:

- *Metric-based:* Huang [1970: multi-parameter family], Oren and Luenberger [1974: self-scaling updates], Osborne and Sun [1988: metric-based self-scaling SR1 update], Zhang and Tewarson [1988: study of variational principles and updates of rank greater than 2], Dennis and More [1977: key survey article], Dennis and Wolkowitz [1990: alternative variational principles], Mifflin and Nazareth [1991: self-complementary update derived from a least prior deviation variational principle], Hoshino [1972: self-complementary update], Luksan[1990: computational study], Al-Baali [1992: computational study].

- *Model-based:* Davidon [1975: rank-2 family that generalizes SR1], Schnabel [1977: comprehensive study of Davidon's 1975 article]. Note that this family of updates was studied in a metric-based setting, but may be more suited to model-based use.

Guo [1993] gives a promising approach that combines metric-based and model-based methods.

2.1.2 Alternative Representations

This is important from a practical standpoint. See Gill, Murray and Pitfield [1972: maintaining reconditioner in triangular form, thus making for efficient solution of associated linear systems and control of their condition number], Brodlie, Gourlay and Greenstadt [1973: product forms of QN updates].

2.2 Truncated

The analogue of truncated Newton methods of Section 1.4. See, for example, Steihaug [1980].

2.3 Limited-Storage

This extension of the QN approach has proved important in practice.

- *Metric-based:* Nocedal [1980: uses the form (2:63) extended to several prior steps (see also Chapter 1, Section 9), along with a suitable recurrence relation to enhance computational efficiency], Liu and Nocedal [1989: computational study of previous algorithm], Siegel [1992: extension to the B-family]. Gilbert and Lemarechal [1989], Zou et al. [1993].

- *Model-based:* See Byrd, Nocedal and Schnabel [1992]. Limited-storage SR1 remains to be fully explored.

2.4 Partitioned

This approach is very important from a practical standpoint. It relies on the fact that problems with sparse Hessians are partially separable, i.e., the function can be expressed as a sum of functions, each involving just a few variables. See Griewank and Toint [1982]

2.5 Sparse

See, for example, Toint [1977], Thapa [1983]. Such methods have not proved themselves in practice.

2.6 Other Models

See, for example, Davidon [1980: conics], Ariyawansa [1990: conics]. The jury is still out on the practical utility of the conic approach as it applies to QN methods for unconstrained nonlinear minimization.

3 CG-Technology

This is again an area where the literature is enormous. As with QN-updates, the subject is delightful from a theoretical standpoint, but the practical utility of many of the proposed variants is unclear. For overviews, see, for example, Stoer [1977], Nazareth and Nocedal [1978], Gilbert and Nocedal [1990].

3.1 Variants

3.1.1 Scaling and Preconditioning.

This is important from a practical standpoint. See, for example, Gill and Murray [1979], Gilbert and Lemarechal [1989], Shanno [1978], Luenberger [1984].

3.1.2 Arbitrary Starting Direction.

See, for example, Beale [1972: extends CG relation so that the initial direction does not have to be along the negative gradient to obtain conjugate directions], Powell [1977: restarting strategy based on Beale's extension].

3.1.3 Inexact Line Searches

See, for example, Nazareth [1979: three-term recurrence that obtains conjugate directions without requiring exact line searches or that initial direction be along the negative gradient direction], Dixon, Ducksbury and Singh [1983: computational study], Dixon [1975], Papadrakis and Ghionis [1986: numerical study showing that the practical merit of many variants is in doubt].

3.2 Limited-Storage

See, for example, Nazareth [1976b], Perry [1976], Shanno [1978: combines the CG-metric with restarting and self-scaling strategies], Nazareth and Nocedal [1982: comprehensive study], Buckley and LeNir [1983: close connection to limited-storage QN], Khoda and Storey [1990: extension of Polak-Ribiere related to SAR/Newton].

3.3 Other models

See, for example, Nazareth and Ariyawansa [1989: conics], Nazareth [1993: trust regions based on conics], Nesterov [1983: algorithm with CG-like features].

4 Cauchy-Technology

See, in particular, Barzilai and Borwein [1988: interesting non-monotonic extension].

BIBLIOGRAPHY

Al-Baali, M. (1992), "An efficient class of quasi-Newton algorithms in the Broyden family", Report No. 9, Department of Electronics, Informatics and Systems, University of Calabria, Calabria, Italy.

Ariyawansa, K.A. (1990), "Deriving collinear scaling algorithms as extensions of quasi-Newton methods and the local convergence of DFP- and BFGS-related collinear scaling algorithms", *Mathematical Programming*, 49, 23-48.

Anderson, H.L. and Davidon, W.C. (1957), "Machine analysis of pion scattering by the maximum likelihood method", *Il Nuovo Cimento*, Serie X, 5, 1238-1255.

Anderson, H.L., Davidon, W.C., Glicksman, M. and Kruse, U.E. (1957), "Scattering of positive pions by hydrogen at 189 Mev", *The Physical Review*, 100, 279-287.

Armijo, L. (1966), "Minimization of functions having Lipschitz-continuous first partial derivatives", *Pacific J. Mathematics*, 16, 1-3.

Avriel, M. (1976), *Nonlinear Programming: Analysis and Methods*, Prentice Hall, Englewood Cliffs, N.J.

Barzilai, J and Borwein, J.M. (1988), "Two-point step size gradient methods", *IMA J. Numerical Analysis*, 8, 141-148.

Bazaraa, M.S., Sherali, H.D. and Shetty, C.M. (1993), *Nonlinear Programming: Theory and Algorithms*, John Wiley, N.Y.

Beale, E.M.L. (1972), "A derivation of conjugate gradients", in *Numerical Methods for Nonlinear Optimization*, F.A. Lootsma (Ed.), Academic Press, N.Y., 39-43.

Brayton, R.K. and Cullum, J. (1979), "An algorithm for minimizing a differentiable function subject to box constraints and errors", *J.O.T.A.*, 29, 521-558.

Brodlie, K.W., Gourlay, A.R. and Greenstadt, J. (1973), "Rank one and two corrections to positive definite matrices expressed in product form", *J.I.M.A.*, 11, 73-82.

Broyden, C.G. (1967), "Quasi-Newton methods and their application to function minimization", *Mathematics of Computation*, 368-381.

Broyden, C.G. (1970), "The convergence of a class of double-rank minimization algorithms", Parts I and II, *J.I.M.A.* , 6, 76-90, 222-236.

Buckley, A. (1978), "Extending the relationship between the conjugate gradient and BFGS algorithms", *Mathematical Programming*, 15, 343-348.

Buckley, A. (1978), "A combined conjugate-gradient quasi-Newton minimization algorithm", *Mathematical Programming*, 15, 200-210.

Buckley, A. and LeNir, A. (1983), "QN-like variable storage conjugate gradients", *Mathematical Programming*, 27, 155-175.

Bunch, J.R. and Parlett, B.N. (1971), "Direct methods for solving symmetric indefinite systems of linear equations", *SIAM J. Numerical Analysis*, 8, 639-655.

Burke, J.V., Moré, J.J. and Toraldo, G. (1990), "Convergence properties of trust region methods for linear and convex constraints", *Mathematical Programming*, 47, 305-336.

Byrd, R.H., Nocedal, J. and Schnabel, R.B. (1992), "Representations of quasi-Newton matrices and their use in limited memory methods", Tech. Report CU-CS-612-92, Department of Computer Science, University of Colorado, Boulder, CO.

Cauchy. A. (1829), "Sur la determination approximative des racines d'une equation algebrique ou transcendante", *Oeuvres Complete (II)*, 4, 573-607. Gauthier-Villars, Paris, 1899.

Conn, A.R., Gould N.I.M. and Toint, Ph. L. (1991), "Convergence of quasi-Newton matrices generated by the symmetric rank one update", *Mathematical Programming*, 50, 177-196.

Coleman, T.F. and Moré, J.J. (1982), "Estimation of sparse Hessian matrices and graph coloring problems", MCS Tech. Memo. TM-4, Argonne National Laboratory, Argonne, IL.

Dahlquist, G. and Bjorck, A. (1974), *Numerical Methods*, Prentice Hall, Englewood Cliffs, N.J.

Davidon, W.C. (1959), "Variable metric method for minimization", Argonne National Laboratory, Report ANL-5990 (Rev.), Argonne, Illinois (reprinted with a new preface in *SIAM J. Optimization*, 1, 1991).

Davidon, W.C. (1968), "Variance algorithm for minimization", *Computer Journal*, 10, 406-410.

Davidon, W.C. (1975), "Optimally conditioned optimization algorithms without line searches", *Mathematical Programming*, 9, 1-30.

Davidon, W.C. (1980), "Conic approximations and collinear scalings for optimizers", *SIAM J. Numerical Analysis*, 17, 268-281.

Davidon, W.C. and Nazareth, J.L. (1977), "OCOPTR - A derivative-free Fortran implementation of Davidon's optimally conditioned method", ANL-AMD Tech. Memo. 303, Applied Mathematics Division, Argonne National Laboratory, Argonne, IL.

Davidon, W.C. and Nazareth, J.L. (1977), "DRVOCR - A Fortran implementation of Davidon's optimally conditioned method", ANL-AMD Tech. Memo. 306, Applied Mathematics Division, Argonne National Laboratory, Argonne, IL.

Davidon, W.C., Mifflin, R.B. and Nazareth, J.L. (1991), "Some comments on notation for quasi-Newton methods", *Optima*, 32, 3-4.

Dembo, R.S., Eisenstat, S.C. and Steihaug, T. (1982), "Inexact Newton Methods", *SIAM J. Numerical Analysis*, 19, 400-408.

Dembo, R.S. and Steihaug, T. (1983), "Truncated Newton algorithms for large-scale optimization", *Mathematical Programming*, 26, 190-212.

Dennis, J.E. and Moré, J.J. (1977), "Quasi-Newton methods: motivation and theory", *SIAM Review*, 19, 46-89.

Dennis, J.E. and Schnabel, R.B. (1981), "A new derivation of symmetric positive definite secant updates", In: *Nonlinear Programming 4*, O.L. Mangasarian, R.R. Meyer and S.M. Robinson, Eds, Academic Press, N.Y.

Dennis, J.E. and Schnabel, R.B. (1983), *Numerical Methods for Unconstrained Optimization and Nonlinear Equations*, Prentice-Hall, N.J.

Dennis, J.E. and Wolkowicz, H. (1990), "Sizing and least change secant updates", CORR Report 90-02, Department of Combinatorics and Optimization, University of Waterloo, Ontario, Canada.

Dixon, L.C.W. (1972), "Quasi-Newton algorithms generate identical points", Parts I and II, *Mathematical Programming*, 2, 383-387; 3, 345-358.

Dixon, L.C.W. (1975), "Conjugate gradient algorithms: quadratic termination without linear searches", *J.I.M.A.*, 15, 9-18.

Dixon, L.C.W., Ducksbury, P.G. and Singh, P. (1983), "A new three term conjugate gradient method", Report No. 130, Numerical Optimization Center, The Hatfield Polytechnic, England.

Fiacco, A.V. and McCormick, G.P. (1968) *Nonlinear Programming: Sequential Unconstrained Minimization Techniques*, John Wiley, N.Y.

Fletcher, R. (1970), "A new approach to variable metric algorithms", *Computer Journal*, 13, 317-322.

Fletcher, R. (1980), *Practical Methods of Optimization, Volume I: Unconstrained Optimization*, John Wiley and Sons. (Second Edition, 1987).

Fletcher, R. and Freeman, T.L. (1977), "A modified Newton method for minimization", *J.O.T.A.*, 23, 357-372.

Fletcher, R. and Powell, M.J.D. (1963), "A rapidly convergent descent method for minimization", *Computer Journal*, 6, 163-168.

Fletcher, R. and Reeves, C. (1964), "Function minimization by conjugate gradients", *Computer Journal*, 7, 149-154.

Gay, D.M. (1981), Computing optimal locally constrained steps", *SIAM J. Sci. Stat. Computing*, 2, 186-197.

Gilbert, J.C. (1992), "Automatic differentiation and iterative processes", *Optimization Methods and Software*, 1, 13-21.

Gilbert, J.C. and Lemarechal, C. (1989), "Some numerical experiments with variable storage quasi-Newton algorithms", *Mathematical Programming*, 45, 407-436.

Gilbert, J.C. and Nocedal, J. (1990), "Global convergence properties of conjugate gradient methods of optimization", Research Report 1268, INRIA, Rocquencourt, France.

Gill, P.E. and Murray, W. (1972), "Quasi-Newton methods for unconstrained optimization", *J. Inst. Math. Applics.*, 9, 91-108.

Gill, P.E. and Murray, W. (1974), "Newton-type methods for unconstrained and linearly constrained optimization", *Mathematical Programming*, 7, 311-350.

Gill, P.E.. and Murray, W. (1979), "Conjugate gradient methods for large-scale nonlinear optimization", Tech. Report SOL 79-15, Department of Operations Research, Stanford University, Stanford, CA.

Gill, P.E., Murray, W. Pickens, W.M. and Wright, M.H. (1979), "The design and structure of a Fortran library for optimization", *ACM Trans. on Math. Software*, 5, 259-283.

Gill, P.E., Murray, W. and Pitfield, P.A. (1972), "The implementation of two revised quasi-Newton algorithms for unconstrained optimization", Report NAC-11, National Physical Labroatory, England.

Gill, P.E., Murray, W. and Wright, M.H.(1981), *Practical Optimization*, Academic Press.

Goldfarb, D. (1970), "A family of variable metric methods derived by variational means", *Mathematics of Computation*, 24, 23-26.

Goldfarb, D. (1980), "Curvilinear path steplength algorithms for minimization which use directions of negative curvature", *Mathematical Programming*, 18, 31-40.

Golub, G.H. and Van Loan, C.F. (1989), *Matrix Computations*, The Johns Hopkins University Press (Second Edition).

Goldstein, A.A. (1962), "Cauchy's method of minimization", *Numerische Mathematik*, 4, 146-150.

Goldstein, A.A. (1965), "On Newton's method", *Numerische Mathematik*, 7, 391-393.

Goldstein, A.A. (1967), *Constructive Real Analysis*, Harper and Row, N.Y.

Greenstadt, J. (1967), "On the relative efficiencies of gradient methods", *Mathematics of Computation*, 21, 360-367.

Greenstadt, J. (1970), "Variations on variable metric methods", *Mathematics of Computation*, 24, 1-22.

Griewank, A. (1989), "On automatic differentiation", in *Mathematical Programming: Recent Developments and Applications*, M. Iri and K. Tanabe (Eds.), Kluwer Academic Publishers, Dordrecht, 83-108.

Griewank, A. and Toint, Ph. L. (1982), "On the unconstrained optimization of partially separable objective functions", in *Nonlinear Optimization, 1981*, M.J.D. Powell (Ed.), Academic Press, London, 301-312.

Guo, J. (1993), "A class of variable metric-secant algorithms for unconstrained minimization", Ph.D. Dissertation, Department of Pure and Applied Mathematics, Washington State University, Pullman, WA.

Hebden, M.D. (1973), "An algorithm for minimization using exact second derivatives", A.E.R.E. Report TP 515, Harwell, England.

Hestenes, M. R. (1980), *Conjugate Direction Methods in Optimization*, Springer Verlag, Berlin.

Hestenes, M.R. and Stiefel, E.L. (1952), "Methods of conjugate gradients for solving linear systems", *J. Res. Nat. Bur. Stds.*, Section B, 49, 409-436.

Hillstrom, K.E. (1976a), "Optimization Routines in AMDLIB", ANL-AMD Tech. Memo. 297, Argonne National Laboratory, Argonne, IL.

Hillstrom, K.E. (1976b), "A simulation test approach to the evaluation and comparison of unconstrained nonlinear optimization algorithms", Report ANL-76-20, Argonne National Laboratory, Argonne, IL.

Hoshino, S. (1972), "A formulation of variable metric methods", *J.I.M.A.*, 10, 394-403.

Hu, Y.F. and Storey, C. (1991), "Motivating quasi-Newton updates by preconditioned conjugate gradient methods", Technical Report A150, Department of Mathematical Sciences, Loughborough University of Technology, England.

Huang, H.Y. (1970), "Unified approach to quadratically convergent algorithms for function minimization", *J.O.T.A.*, 5, 405-423.

Kahaner, D., Moler, C. and Nash, S. (1989), *Numerical Methods and Software*, Prentice Hall, Englewood Cliffs, N.J.

Khalfan, H., Byrd, R.H. and Schnabel, R.B. (1993), "A theoretical and experimental study of the symmetric rank one update", *SIAM J. Optimization*, 3, 1-24.

Khoda, K.M. and Storey C. (1990), "A generalized Polak-Ribiere algorithm", Technical Report A128, Department of Mathematical Sciences, Loughborough University of Technology, England.

Khoda, K.M., Liu, Y. and Storey C. (1992), "Optimized software for a generalized Polak-Ribiere algorithm in unconstrained optimization", Technical Report

A156, Department of Mathematical Sciences, Loughborough University of Technology, England.

Levenberg, K. (1944), "A method for the solution of certain problems in least squares", *Quarterly of Applied Mathematics*, 2, 164-168.

Liu, D.C. and Nocedal, J. (1989), "On the limited memory BFGS method for large-scale optimization", *Mathematical Programming*, 45, 503-528.

Luenberger, D.G. (1984), *Linear and Nonlinear Programming*, Addison-Wesley (Second Edition).

Luksan, L. (1990), "Computational experience with improved variable metric methods for unconstrained optimization", *Kybernetika*, 26, 415-431.

Lyness, J.N. (1979), "A benchmark experiment for minimization algorithms", *Mathematics of Computation*, 33, 249-264.

Marquardt, D.W. (1963), "An algorithm for least squares estimation of nonlinear parameters", *SIAM J. Applied Mathematics*, 11, 431-441.

Mifflin, R.B. and Nazareth, J.L. (1991), "The least prior deviation quasi-Newton update", Technical Report TR 91-1, Department of Pure and Applied Mathematics, Washington State University (to appear in *Mathematical Programming*).

Moré, J.J. (1983), "Recent developments in algorithms and software for trust region methods", in *Mathematical Programming: The State of the Art, Bonn 1982*, A. Bachem, M. Grotschel and B. Korte (Eds.), Springer Verlag, Berlin, 258-287.

Moré, J.J., Garbow, B.S. and Hillstrom, K.E. (1980), "User's Guide to Minpack-1", Report ANL-80-74, Argonne National Laboratory, Argonne, IL.

Moré, J.J., Garbow, B.S. and Hillstrom, K.E. (1981), "Testing unconstrained optimization software", *ACM Trans. on Mathematical Software*, 7, 17-41.

More, J.J. and Sorensen, D.C. (1982), "Newton's method", Report ANL-82-8, Argonne National Labortory, Argonne, IL.

Moré, J.J. and Sorensen, D.C. (1983), "Computing a trust region step", *SIAM J. Scientific and Statistical Computing*, 4, 553-572.

Moré, J.J. and D.J. Thuente (1990), "On line search algorithms with guaranteed sufficient decrease", MCS Division Preprint MCS-P153-0590, Argonne National Laboratory, Argonne, IL.

Nash, S.G. (1985), "Preconditioning of truncated-Newton methods", *SIAM J. Scientific and Statistical Computing*, 6, 599-616.

Nash, S.G. and Nocedal, J. (1989), "A numerical study of the limited memory BFGS method and the truncated-Newton method for large scale optimization", Tech. Report AC-89-08, EECS Department, Northwestern University, Evanston, IL.

Nazareth, J.L. (1975), "A hybrid least squares method", Tech. Memo. ANL-AMD 254, Argonne National Laboratory, Argonne, IL.

Nazareth, J.L. (1976a), "Generation of conjugate directions for unconstrained minimization without derivatives", *Mathematics of Computation*, 30, 115-131.

Nazareth, J.L. (1976b), "A relationship between the BFGS and conjugate gradient algorithms", Tech. Memo. ANL-AMD 282, Argonne National Laboratory. Presented at the SIAM-SIGNUM Fall 1975 Meeting, San Francisco, CA. (Also appears in *SIAM J. Numerical Analysis*, 16, 794-800, 1979.)

Nazareth, J.L. (1976c), "On Davidon's optimally conditioned algorithm for unconstrained optimization", Tech. Memo. ANL-AMD 283, Argonne National Laboratory, Argonne, IL.

Nazareth, J.L. (1977a), "Unified approach to unconstrained minimization via basic matrix factorizations", *Linear Algebra and its Applications*, 17, 197-232.

Nazareth, J.L. (1977b), "MINKIT - An optimization system", ANL-AMD Tech. Memo. 305, Argonne National Laboratory, Argonne, IL.

Nazareth, J.L. (1980), "Some recent approaches to solving large-residual nonlinear least squares problems", *SIAM Review*, 22, 1-11.

Nazareth, J.L. (1979), "A conjugate direction algorithm for unconstrained minimization without line searches", *J.O.T.A.*, 23, 373-387.

Nazareth, J.L. (1984), "An alternative variational principle for variable metric updating", *Mathematical Programming*, 30, 99-104.

Nazareth, J.L. (1985), "Hierarchical implementation of optimization methods", in *Numerical Optimization, 1984*, P. Boggs, R.H. Byrd and R.B. Schnabel (Eds.), SIAM, Philadelphia, 199-210.

Nazareth, J.L. (1986a), "Analogues of Dixon's and Powell's theorems for unconstrained minimization with inexact line searches", *SIAM J. Numerical Analysis*, 23, 170-177.

Nazareth, J.L. (1986b), "Conjugate gradient algorithms less dependent on conjugacy", *SIAM Review*, 28, 501-511.

Nazareth, J.L. (1986c), "The method of successive affine reduction for nonlinear minimization", *Mathematical Programming*, 35, 97-109.

Nazareth, J.L. (1986d), "An algorithm based upon successive affine reduction and Newton's method", In: *Proceedings of the Seventh INRIA International Conference on Computing Methods in Applied Science and Engineering*, (Versailles, France), R. Glowinski and J-L. Lions, Eds., North-Holland, 641-646.

Nazareth, J.L. (1992), "The Newton and Cauchy perspectives on computational nonlinear optimization", Presented at *Three Decades of Numerical Linear Algebra at Berkeley*, A Conference in Honour of the 60'th Birthdays of Profs. B.N. Parlett and W. Kahan, Univeristy of California, Berkeley (October, 1992).

Nazareth, J.L. (1993), "Trust regions based on conic functions for linear and nonlinear programming", *J. Numerical Linear Algebra and Applications*, to appear.

Nazareth, J. L. and Ariyawansa, K. (1989), "On accelerating Newton's method based on a conic model", *Information Processing Letters*, 30, 277-281.

Nazareth, J.L. and Nocedal, J. (1978), "A study of conjugate gradient methods", Tech. Report SOL 78-29, Department of Operations Research, Stanford University, Stanford, CA.

Nazareth, J.L. and Nocedal, J. (1982), "Conjugate direction methods with variable storage", *Mathematical Programming*, 23, 326-340.

Nemirovsky, A.S. and Yudin, D.B. (1983), *Problem Complexity and Method Efficiency*, John Wiley, NY.

Nesterov, Y.E. (1983), "A method of solving a convex programming problem with convergence rate $O(1/k^2)$", *Soviet Mathematics Doklady*, 27, 372-376.

Nocedal, J. (1980), "Updating quasi-Newton matrices with limited storage", *Mathematics of Computation*, 35, 773-782.

Nocedal, J. (1991), "Theory of algorithms for unconstrained optimization", *Acta Numerica*, 1, 199-242.

Oren, S.S (1974), "On the selection of parameters in self-scaling variable metric algorithms", *Mathematical Programming*, 7, 351-367.

Oren, S.S. and Luenberger, D.C. (1974), "Self-scaling variable metric (SSVM) algorithms. Part 1. Criteria and sufficient conditions for scaling a class of algorithms, Part 2: Implementation and experiments. *Management Science*, 20, 845-862. 863-874.

Ortega, J.M. and Rheinboldt, W.C. (1970), *Iterative Solution of Nonlinear Equations in Several Variables*, Academic Press.

Osborne, M.R. and Sun, L.P. (1988), "A nw approach to the symmetric rank one updating algorithm", REport No, NMO/01, School of Mathematics, Australian National University.

Parlett, B.N. (1980), *The Symmetric Eigenvalue Problem*, Prentice-Hall, englewood Cliffs, N.J.

Papadrakakis, M. and Ghionis, P (1986), "Conjugate gradient algorithms in nonlinear structural analysis", *Computer Methods in Applied Mechanics and Engineering*, 59, 11-27.

Pearson, J.D. (1969), "Variable metric methods of minimization", *Computer Journal*, 12, 171-178.

Perry, A. (1976), "A modified CG algorithm", Discussion Paper 229, Center for Mathematical Studies in Economics and Management Science, Northwestern

University, Evanston, IL. (also appears in *Operations Research*, 26, 1073-1078, 1978).

Polak, E. (1971), *Computational Methods in Optimization: A Unified Approach*, Academic Press.

Polak, E. and Ribiere, G. (1969), "Note sur la convergence de methode de directions conjuguees", *Revue Francaise d'Informatique et de Recherche Operationnelle*, 16, 35-43.

Powell, M.J.D. (1972), "Unconstrained minimization and extensions for constraints", A.E.R.E. Report TP 495, Harwell, England.

Powell, M.J.D. (1976), "Some global convergence properties of a variable metric algorithm for minimization without exact line searches", In: *Nonlinear Programming, SIAM-AMS Proceedings Vol. IX*, R.W. Cottle and C.E. Lemke, Eds., SIAM, Philadelphia.

Powell, M.J.D. (1977), "Restart procedures for the conjugate gradient method", *Mathematical Programming*, 12, 241-254.

Powell, M.J.D. (1984), "On the global convergence of trust region algorithms for unconstrained minimization", *Mathematical Programming*, 29, 297-303.

Powell, M.J.D. (1986), "How bad are the BFGS and DFP methods when the objective function is quadratic", *Mathematical Programming*, 34, 34-47.

Powell, M.J.D. and Ph. L. Toint (1979), "On the estimation of sparse Hessian matrices", *SIAM J. Numerical Analysis*, 16, 1060-1074.

Press, W.H., Vetterling, W.T., Teukolsky, S.A. and Flannery, B.P. (1992), *Numerical Recipes in Fortran: The Art of Scientific Computing*, (Second Edition), Cambridge University Press, Cambridge, England.

Reinsch, C.H. (1971), "Smoothing by spline functions II", *Numerische Mathematik*, 16, 451-454.

Rice, J.R. (1971), "The challenge of mathematical software", in *Mathematical Software*, J. Rice (Ed.), Academic Press, N.Y., 27-41.

Rudin, W. (1976), *Principles of Mathematical Analysis*, McGraw-Hill (Third Edition).

Schnabel, R.B. (1977), "Analysing and improving quasi-Newton methods for unconstrained optimization", Tech. Report TR-77-320, Department of Computer Science, Cornell University, Ithaca, N.Y.

Schnabel, R.B. (1993), "Tensor method software packages for nonlinear equations and least squares, and unconstrained optimization", *SIAG/OPT Views-and-News*, 2, 3-5.

Schnabel, R.B. and Chow, T. (1991), "Tensor methods for unconstrained optimization using second derivatives", *SIAM J. Optimization*, 1, 293-315.

Shanno, D.F. (1970), "Conditioning of quasi-Newton methods for function minimization", *Mathematics of Computation*, 24, 647-657.

Shanno, D.F. (1978), "Conjugate gradient methods with inexact searches", *Mathematics of Operations Research*, 3, 244-256.

Siegel, D. (1992), "Implementing and modifying Broyden class updates for large-scale optimization", Report DAMTP 1992/NA12, University of Cambridge, England.

Smith, B.T., Boyle, J.M. and Cody, W.J. (1974), "The NATS approach to quality software", in *Proceedings of IMA Conference on Software for Numerical Mathematics*, J. Evans (Ed.), Academic Press, N.Y., 393-405.

Smith, B. and Nazareth, J.L. (1993), "Metric-based SR1 updates", (in preparation).

Sorensen, D.C. (1982a), "Newton's method with a model trust region modification", *SIAM J. Numerical Analysis*, 19, 409-426.

Sorensen, D.C. (1982b), "Trust region methods for unconstrained optimization", in *Nonlinear Optimization 1981*, M.J.D. Powell (Ed.), Academic Press, N.Y.

Sorensen, D.C. (1977), "Updating the symmetric indefinite factorization with applications to a modified Newton's method", Ph.D. thesis, University of California, San Diego.

Steihaug, T. (1980), "Quasi-Newton methods for large scale nonlinear problems", Working Paper No. 49, School of Organization and Management, Yale University, New Haven, CT.

Stoer, J. (1977), "On the relation between quadratic termination and convergence properties of minimization algorithms, Part I, Theory", *Numerische Mathematik*, 28, 343-366.

Stoer, J. (1983), "Solution of large linear systems of equations by conjugate gradient type methods", in *Mathematical Programming: The State of the Art, Bonn 1982*, A. Bachem, M. Grotschel and B. Korte (Eds.), Springer Verlag, Berlin, 540-565.

Thapa, M. N. (1983), "Optimization of unconstrained functions with sparse Hessian matrices: quasi-Newton methods", *Mathematical Programming*, 25, 158-182.

Toint, Ph. L. (1977), "On sparse and symmetric matrix updating subject to a linear constraint", *Mathematics of Computation*, 32, 954-961.

Trefethen, L. N. (1992), "The definition of numerical analysis", *SIAM News*, 25, p.6.

Watkins, D. (1991), *Fundamentals of Matrix Computations*, John Wiley.

Wilkinson, J.H (1965), *The Algebraic Eigenvalue Problem*, Clarendon Press.

Wolfe, P. (1969), "Convergence conditions for ascent methods", *SIAM Review*, 11, 226-235.

Wolfe, P. (1971), "Convergence conditions for ascent methods II: some corrections", *SIAM Review*, 13, 185-188.

Ypma, T.J. (1992), "Historical development of the Newton-Raphson method", Department of Mathematics, Western Washington University, Bellingham, WA 98225-9063, USA (preprint).

Zangwill, W.I. (1969), *Nonlinear Programming: A Unified Approach*, Prentice Hall, Englewood Cliffs, N.J.

Zhang, Y. and Tewarson, R.P. (1987), "Least-change secant updates for Cholesky factors subject to the nonlinear quasi-Newton condition", *IMA J. Numerical Analysis*, 7, 509-521.

Zhang, Y. and Tewarson, R.P. (1988), "Quasi-Newton algorithms with updates from the preconvex part of Broyden's family", *IMA J. Numerical Analysis*, 8, 487-509.

Zou, X, Navon, I.M., Berger, M., Phue, K.H., Schlick, T. and LeDimet F.X. (1993), "Numerical experience with limited-memory quasi-Newton and truncated Newton methods", *SIAM J. Optimization*, to appear.

Zoutendijk, G. (1970), "Nonlinear programming computational methods", in *Integer and Nonlinear Programming*, J. Abadie (Ed.), North-Holland, Amsterdam, 37-86.

Printing: Weihert-Druck GmbH, Darmstadt
Binding: Buchbinderei Schäffer, Grünstadt

Lecture Notes in Computer Science

For information about Vols. 1–699
please contact your bookseller or Springer-Verlag

Vol. 736: R. L. Grossman, A. Nerode, A. P. Ravn, H. Rischel (Eds.), Hybrid Systems. VIII, 474 pages. 1993.

Vol. 737: J. Calmet, J. A. Campbell (Eds.), Artificial Intelligence and Symbolic Mathematical Computing. Proceedings, 1992. VIII, 305 pages. 1993.

Vol. 738: M. Weber, M. Simons, Ch. Lafontaine, The Generic Development Language Deva. XI, 246 pages. 1993.

Vol. 739: H. Imai, R. L. Rivest, T. Matsumoto (Eds.), Advances in Cryptology – ASIACRYPT '91. X, 499 pages. 1993.

Vol. 740: E. F. Brickell (Ed.), Advances in Cryptology – CRYPTO '92. Proceedings, 1992. X, 593 pages. 1993.

Vol. 741: B. Preneel, R. Govaerts, J. Vandewalle (Eds.), Computer Security and Industrial Cryptography. Proceedings, 1991. VIII, 275 pages. 1993.

Vol. 742: S. Nishio, A. Yonezawa (Eds.), Object Technologies for Advanced Software. Proceedings, 1993. X, 543 pages. 1993.

Vol. 743: S. Doshita, K. Furukawa, K. P. Jantke, T. Nishida (Eds.), Algorithmic Learning Theory. Proceedings, 1992. X, 260 pages. 1993. (Subseries LNAI)

Vol. 744: K. P. Jantke, T. Yokomori, S. Kobayashi, E. Tomita (Eds.), Algorithmic Learning Theory. Proceedings, 1993. XI, 423 pages. 1993. (Subseries LNAI)

Vol. 745: V. Roberto (Ed.), Intelligent Perceptual Systems. VIII, 378 pages. 1993. (Subseries LNAI)

Vol. 746: A. S. Tanguiane, Artificial Perception and Music Recognition. XV, 210 pages. 1993. (Subseries LNAI).

Vol. 747: M. Clarke, R. Kruse, S. Moral (Eds.), Symbolic and Quantitative Approaches to Reasoning and Uncertainty. Proceedings, 1993. X, 390 pages. 1993.

Vol. 748: R. H. Halstead Jr., T. Ito (Eds.), Parallel Symbolic Computing: Languages, Systems, and Applications. Proceedings, 1992. X, 419 pages. 1993.

Vol. 749: P. A. Fritzson (Ed.), Automated and Algorithmic Debugging. Proceedings, 1993. VIII, 369 pages. 1993.

Vol. 750: J. L. Díaz-Herrera (Ed.), Software Engineering Education. Proceedings, 1994. XII, 601 pages. 1994.

Vol. 751: B. Jähne, Spatio-Temporal Image Processing. XII, 208 pages. 1993.

Vol. 752: T. W. Finin, C. K. Nicholas, Y. Yesha (Eds.), Information and Knowledge Management. Proceedings, 1992. VII, 142 pages. 1993.

Vol. 753: L. J. Bass, J. Gornostaev, C. Unger (Eds.), Human-Computer Interaction. Proceedings, 1993. X, 388 pages. 1993.

Vol. 754: H. D. Pfeiffer, T. E. Nagle (Eds.), Conceptual Structures: Theory and Implementation. Proceedings, 1992. IX, 327 pages. 1993. (Subseries LNAI).

Vol. 755: B. Möller, H. Partsch, S. Schuman (Eds.), Formal Program Development. Proceedings. VII, 371 pages. 1993.

Vol. 756: J. Pieprzyk, B. Sadeghiyan, Design of Hashing Algorithms. XV, 194 pages. 1993.

Vol. 757: U. Banerjee, D. Gelernter, A. Nicolau, D. Padua (Eds.), Languages and Compilers for Parallel Computing. Proceedings, 1992. X, 576 pages. 1993.

Vol. 758: M. Teillaud, Towards Dynamic Randomized Algorithms in Computational Geometry. IX, 157 pages. 1993.

Vol. 759: N. R. Adam, B. K. Bhargava (Eds.), Advanced Database Systems. XV, 451 pages. 1993.

Vol. 760: S. Ceri, K. Tanaka, S. Tsur (Eds.), Deductive and Object-Oriented Databases. Proceedings, 1993. XII, 488 pages. 1993.

Vol. 761: R. K. Shyamasundar (Ed.), Foundations of Software Technology and Theoretical Computer Science. Proceedings, 1993. XIV, 456 pages. 1993.

Vol. 762: K. W. Ng, P. Raghavan, N. V. Balasubramanian, F. Y. L. Chin (Eds.), Algorithms and Computation. Proceedings, 1993. XIII, 542 pages. 1993.

Vol. 763: F. Pichler, R. Moreno Díaz (Eds.), Computer Aided Systems Theory – EUROCAST '93. Proceedings, 1993. IX, 451 pages. 1994.

Vol. 764: G. Wagner, Vivid Logic. XII, 148 pages. 1994. (Subseries LNAI).

Vol. 765: T. Helleseth (Ed.), Advances in Cryptology – EUROCRYPT '93. Proceedings, 1993. X, 467 pages. 1994.

Vol. 766: P. R. Van Loocke, The Dynamics of Concepts. XI, 340 pages. 1994. (Subseries LNAI).

Vol. 767: M. Gogolla, An Extended Entity-Relationship Model. X, 136 pages. 1994.

Vol. 768: U. Banerjee, D. Gelernter, A. Nicolau, D. Padua (Eds.), Languages and Compilers for Parallel Computing. Proceedings, 1993. XI, 655 pages. 1994.

Vol. 769: J. L. Nazareth, The Newton-Cauchy Framework. XII, 101 pages. 1994.

Vol. 770: P. Haddawy (Representing Plans Under Uncertainty. X, 129 pages. 1994. (Subseries LNAI).

Vol. 771: G. Tomas, C. W. Ueberhuber, Visualization of Scientific Parallel Programs. XI, 310 pages. 1994.

Vol. 772: B. C. Warboys (Ed.),Software Process Technology. Proceedings, 1994. IX, 275 pages. 1994.

Vol. 773: D. R. Stinson (Ed.), Advances in Cryptology – CRYPTO '93. Proceedings, 1993. X, 492 pages. 1994.

Vol. 774: M. Banatre, P. A. Lee (Eds.), Hardware and Software Architectures for Fault Tolerance. XIII, 311 pages. 1994.

Vol. 775: P. Enjalbert, E. W. Mayr, K. W. Wagner (Eds.), STACS 94. Proceedings, 1994. XIV, 782 pages. 1994.